Symbols of Infection
The Story of Immunity

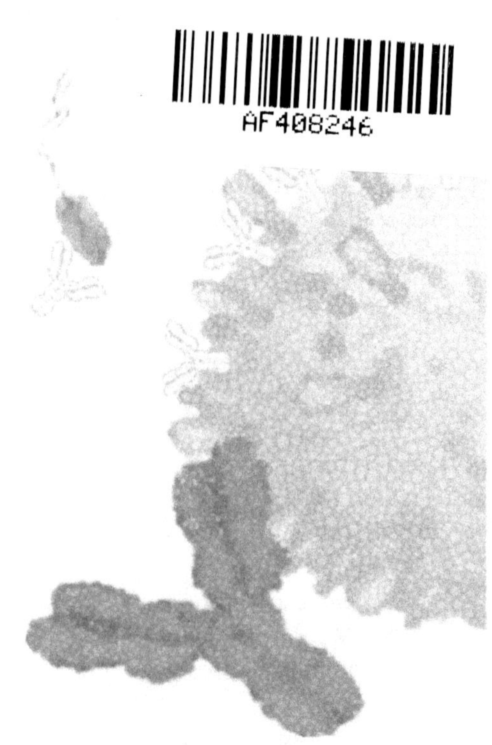

From the author/editor of:
"Endotoxin Detection and Control in
Pharma, Limulus, & Mammalian Systems"
(Springer Nature, 2019), Kevin L. Williams

ISBN: 9798632650717

"The capacity to blunder slightly is the real marvel of DNA. Without this special attribute, we would still be anaerobic bacteria and there would be no music."
Lewis Thomas in The Medusa and the Snail: More Notes of a Biology Watcher.

 **_Symbols of Infection_**

Preface

Chapter 1. Microbial & evolutionary backdrop:
PAMPs & PRRs

Chapter 2. Mammalian innate immunity

Chapter 3. Workhorse structures & function of adaptive Immunity

13. How has the detection architecture developed, ancient to modern, as used in *adaptive* immunity?
14. What is the B cell structure and routine function?
15. How is antibody diversity generated?
16. How does the same membrane-bound BCR become a secreted, soluble antibody? And what is class-switching?
17. How do cells of the organism *"know"* who is self and who is non-self? That is, how is central tolerance established?
18. What is the TCR structure and routine function?
19. What is the relationship between HLA, MHC and TCR and the collective function?

Chapter 4. Adaptive immunity uses innate as a fail-safe

20. What are surface markers and receptors?
21. Where does the "second signal" come from?
22. What are CD4$^+$ and CD8$^+$ T cells?
23. How is long-term memory established?
24. What are the fail-safe practices that prevent these complex systems from attacking themselves (autoimmunity)? And what happens when this breaks down?
25. Why is co-stimulation/co-inhibition a powerful therapeutic paradigm?

Chapter 5. The biologics revolution

26. What is a biologic drug?
27. What is the difference between a biologic developed in 1900 and 2000?
28. What is the central dogma of the *"recombinant revolution"*?
    28.1. Recombinant technology
    28.2 What are monoclonal antibodies and how are they produced?
29. What are some of the newest forms of biologic therapeutics?
    29.1 Immune checkpoint inhibitors
    29.2 CAR-T therapy
    29.3 CD47–"Don't eat me" surface marker and competitive commercial development efforts
30. If the sky is the limit in biotechnology, what is the sky?
    30.1 Modification of natural molecules to make new entities
    30.2 Anti-sense therapy
    30.3 Gene therapy
    30.4 Monster indications and difficult disease targets

Preface

The first thing one is likely to notice about this book is that it is small given the vast topic, the story of immunity. This was the deepest desire in producing this book, to whittle the subject down to the best concepts that cover the most important topics needed to frame the subject. I wrote it because I couldn't find one like it. Here the *story of immunity* is told from the vantage of the single-celled versus multicellular struggle, or arms race, to explore the evolving interface between the two.

This extreme focus contrasts other books on the subject where the major concepts of immunology have been laboriously detailed, but remain painfully complex. It is an effort to paint the most modern concepts with a broad brush in as linear a fashion as is practical (ancient to modern). As can be seen in the table of contents, each chapter answers the most fundamental questions in an overarching manner.

Many of these questions have only been answered recently, but the answers have snowballed in complexity and lead to a revolution in the knowledge of immunology that has been paralleled by advances in therapeutic treatments including recombinant molecule manufacture (biologics) as used to modulate wayward immune responses including in cancer, infection and autoimmunity.

This book presents a survey of many intricate modern concepts of evolution, the evolutionary development of immunity and the "biotech revolution" in an easy to digest format. It is not a science textbook, although it seeks to lay down the basic frame for additional complex and accumulating minutiae of immunology from a microbiology perspective.

If we can get the "big stuff" right, the <u>overarching</u> patterns by which immunity proceeds in nature, then it makes it much easier to fill in the myriad of details that are really just specific puzzle pieces of the already outlined picture. The graphic below shows the story in crude fashion via the "workhorse" structures of innate and adaptive immunity with associated microbial structures. The Q&A format allows the treatment of the topics in a way that is more "*story-like*". The linear story begins with the advent of multicellular life as derived from single celled organisms and ends with a reach to the "sky" via man's current efforts to use biotechnology to solve medical conundrums wound into the metazoan body plan.

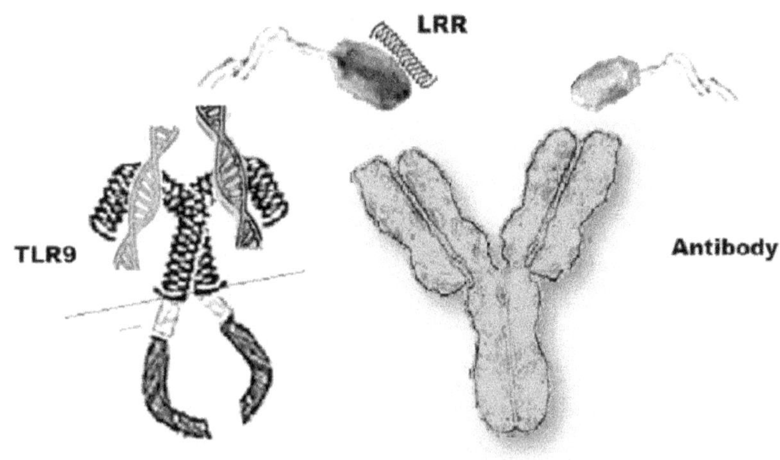

The on-going struggle at the interface of microbes (PAMPs/Antigens) and man (Innate PRRs: TLRs / Adaptive: Antibodies).

Chapter 1. Microbial & evolutionary backdrop: PAMPs & PRRs

PAMPs are "pathogen associated molecular patterns" that are associated with prokaryotic cells, not just pathogens. Some have renamed these MAMPs or microbe associated molecular patterns. PPRs are "pattern recognition receptors" including TLRs (Toll-like receptors) that metazoans (multicellular animals) use to detect and subsequently counter PAMPs.

1. What was the first great evolutionary divide from which much of current disease causation originates?

The defining problem encountered by multicellular (or metazoan) forms that persists to this day with every individual is associated with its internal, cellular efforts to prevent infection by the original ubiquitous unicellular life forms, particularly bacterial and viral types, even as metazoans interact with and contain them (as unwelcomed invaders, symbionts or parasites). Distinguishing self from non-self on a cellular level and battling the myriad of onslaughts and evolving weapons of single-celled invaders occupies an inordinate amount of genetic effort and metabolic energy for metazoans and is not always successful despite cumulative evolutionary efforts. Buetler expressed that infection exerts the highest evolutionary pressure on man given that childhood disease precludes reproduction.[1] The added complexity associated with this "war" adds to the propensity for structures and interactions to "go wrong" in the form of cancer, autoimmunity, and infectious response.

This battle or **"arms race"** as it has been called is the defining fact of life as the interaction and interface of the metazoan and microbial dynamic is wound into the basic paradigms of genetics, infectious disease and evolution. See the graphic below. When you think about it, virtually all disease has been found to have some causation associated with the protective dynamic that is a necessity of genetic change interaction with infectious agents. Infection is the most obvious result of this interaction but also cancer occurs via oncogenes of viral origin and autoimmune disease is a result of the necessity of possessing and maintaining such complex anti-infective architecture.

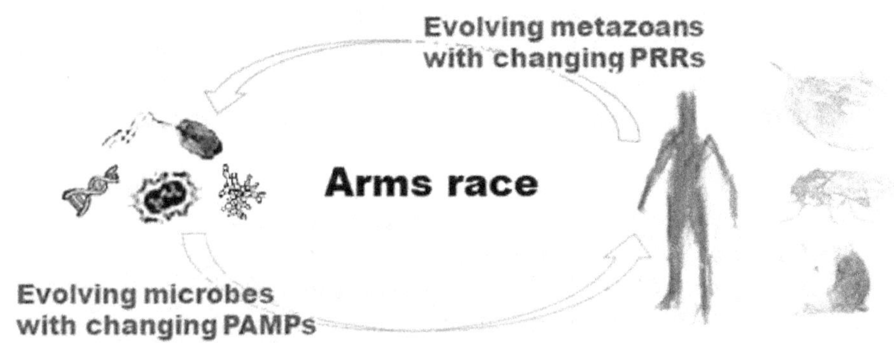

Sepsis is associated with remnants of bacterial infection even in the absence of living microbes. Vaccination has been able to lessen the pain of some of the worst microbial offenders in polio, tetanus, smallpox, rabies, and in some cases typhoid and cholera. And foodborne disease prevention has also gained from basic techniques including pasteurization, refrigeration, water treatment and thorough cooking prior to packaging to prevent toxins including botulism.

Some have theorized that cancer is a reversion to a previously single-celled dynamic in metazoans.

> A decade ago, a radical theory of cancer emerged: that this plague of multicellular organisms arises when their cells rewind the evolutionary clock and revert to acting like unicellular life. Recently, David Goode, a computational cancer biologist at the Peter MacCallum Cancer Centre in Melbourne, Australia, and colleagues have found evidence to support that idea. They examined gene expression in seven types of solid tumors—including breast, stomach, and liver cancers—and traced the ancestry of the active genes they found. Genes that date all the way back to early single-celled eukaryotic organisms were revved up, Goode's team reported last year in the *Proceedings of the National Academy of Sciences*. In contrast, genes unique to many-celled animals had gone quiet.[2]

The entirety of the remaining book weaves an overview of the on-going dynamics of the microbe and metazoan interface.

2. *What was the first known multicellular organism?*

According to Keim,

> Single-celled organisms emerged from the primordial soup about 3.4 billion years ago. Almost immediately, some gathered in mats. But it was another 1.4 billion years before the first truly multicellular organism, called *Grypania spiralis*, appears in the fossil record.
>
> *Grypania* may have been either a bacterial colony or a eukaryote – an organism with specialized cells, enclosed in a membrane. Whatever *Grypania* was, it was one of the few known examples of complex life until about 550 million years ago, when the fossil record explodes in diversity.
>
> The newly described fossils, which have yet to be given a species name, make *Grypania* less solitary. They lived at roughly the same time – *Grypania* in what is now the northern United States, the new fossils in Gabon (west Africa). By raising the possibility that multicellularity was a trend rather than an aberration, they also hint at an answer to the question of *why* complex life evolved, not just when.
>
> Just a few million years before *Grypania* and the newly discovered fossils appear in the fossil record, Earth experienced what's called the Great Oxidation Event. The sudden evolution of photosynthesizing bacteria radically changed Earth's atmosphere, kick-starting its transformation from nearly oxygen-free into today's breathable air. [3]

Libby et al. have studied various ways in which multicellularity may have evolved. The graphic from their experiments below, though using a yeast (a single-celled organism), shows simplistically how successive generations of cellular activity could "unfold" into a differentiated "being". The unfolding of today's complex multicellular animals via Hox genes (Homeobox) is based upon forming gradients of segmentation (in arthropods in literal segments and in mammals in less obvious vertebral segments) to lay out what genes will be expressed where to make what. Therefore, in animals (invertebrates and vertebrates) the generations of multicellularity are directed by a governing system that has only recently been elaborated. The unfolding of the complex multicellular body plan of mammals (fly, mouse and man) is shown below the yeast figure.

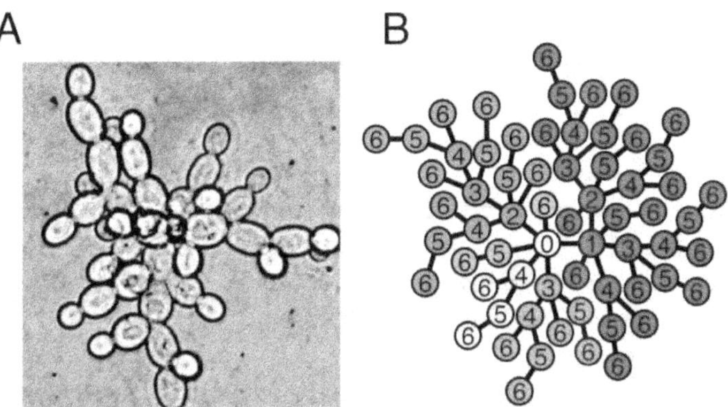

Groups as trees. **A)** Photograph of a cross-section of the yeast snowflake phenotype shows the branching morphology. **B)** Simulated group growth from a single-cell (*Node 0*) after 6 rounds of cell reproduction (generations). The different colors represent different branches emanating from *Node 0*. The numbers inside nodes represent the generation of their birth.

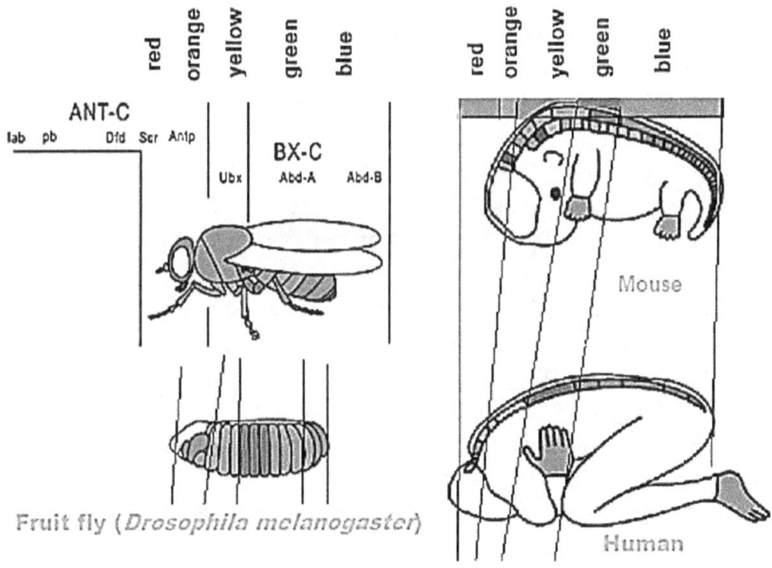

Hox genes are shared by primitive to complex multicellular animals[4,5] as a means of determining the multicellular body plan by establishing the dorso-ventral axis. Truncated and derived from Heuber et al. (CC 4.0)[6] Changes made: lines, greyscale and color names added.

Jack and Du Pasqier describe a "social amoeba", *Dictyostelium discoideum,* that lives a single-cell lifestyle but after consuming all available bacteria, they crowd together to form a "slug" (100,000 cells) which moves around together searching for more food. If it can't find more then it attaches to a substrate and forms a "stalk with a fruiting body" and sends out spores to ensure the survival of the species.

3.   *What were the early foundations of "germ theory"?*
As early as the 1670's, Anton van Leuwenhoek discovered *"animalcules"* in a drop of rain water in the eye of his little homemade microscope, however, he did not conceive of the microbial roots of disease or microbial role in decay, food spoilage, or infection.

> I discovered living creatures in rain, which had stood but a few days in a new tub that was painted blue within. This observation provoked me to investigate this water more narrowly; and especially because these little animals were, to my eye, more than ten thousand times smaller than the animalcule that Jan Swammerdam has portrayed, and called by the name of water flea, or water louse, which you can see alive and moving in water with the bare eye.[7]

Pasteur (1850-1880's), Koch (1880's) and Lister (1860-1880's), elaborated the microbial world and problems of infection and disinfection and Pasteur even developed microbes as tools for fermentation and vaccination. This very basic foundation was later referred to as "germ theory". A bit later, the realization that contaminants also come from artifacts or pieces or secretions (toxins) of microorganisms including endotoxin and exotoxins etc. was also demonstrated and became a cornerstone of vaccination and sterile drug manufacture.

Pfeiffer is credited with ascribing the deleterious non-infectious aspect of sub-bacterial substances (PAMPs), in this case endotoxin, as described below:

> Two crucial observations, and their corresponding ingenious interpretation, led Richard Pfeiffer to formulate the endotoxin hypothesis. He observed that guinea pigs actually died after the application of an appropriate quantity of *Vibrio cholerae* bacteria either together with an immune serum directed against the Vibrios or when injected into previously actively immunised animals. However, after some time, in neither case, no living Vibrios could be detected in the abdominal cavity This result, which caused worldwide sensation, was termed "the Pfeiffer reaction". In fact, this phenomenon violated one of Koch's postulates (recovery of the infectious agent), and Pfeiffer found himself in a dilemma to come up with a plausible explanation...

> Pfeiffer then made a second crucial observation: he showed that the heat killing of cholera bacteria did not weaken their toxic potential distinguishing it from the known (e.g. diploteria) toxins which are heat-labile. Since dead

cells had lost their cytoplasm, Pfeiffer concluded that the toxin was not present in the cytoplasm but, rather, had to be associated with the "body substance", as he called it, i.e., with the insoluble part of the bacterium. These insightful considerations led Pfeiffer to formulate the concept of endotoxin as a poison which, during microbial life, was firmly bound to the bacterial cell and only released post mortem to evolve its pathogenic effects.[8]

The body responds to infectious microbial agents or via the recognition of various pathogen-associated molecular patterns (PAMPs) or molecular structures or "symbols of infection" via pattern recognition receptors (PRRs). Therefore, the non-infectious response mimics the early infectious response even in the absence of living invaders. Amongst the strongest "symbols" of infection include endotoxins and super antigens (SAG). Both are very potent yet only endotoxin is ubiquitous and resistant to degradation via heat.

The basic tenants of pharmaceutical microbiological control has for decades revolved around the detection of these symbols. The longstanding Rabbit Pyrogen Test (RPT) was the first quality control test used to detect non-living contaminants based upon the occurrence of fever after the injection of sterile solutions. The use of fever (a systemic response) as a basis of detection (USP 1942) was largely replaced by the *Limulus* amebocyte test some decades later (beginning in the late 1970's). The mechanism of action used the formation of coagulation in the horseshoe crab's hemolymph to detect endotoxin as a pyrogen causing PAMP. Given that endotoxin is the predominate PAMP able to occur in sterile water based drug manufacturing, it has endured to this day as the most practical and useful test.

The basic outline of "germ theory" led fairly quickly to improvements in basic living conditions due to sanitation, disinfection and the reduction of infectious disease devastation via vaccination, and, most recently via an increasing understanding of disease causation, the deciphering of the microbial genome (as a model) and, more and more, through sophisticated characterization and annotation of the human genome.

4.   *What are the ancient PAMPs that have threatened multicellular organisms since their inception?*

A PAMP in simplistic terms is a piece or artifact (bits and pieces) or secretion of a bacterial, viral, protozoan or fungal cell. Microbial cell walls of all types contain many such molecular "symbols" or markers including endotoxin (Gram Neg.), peptidoglycan (Gram pos.), beta-glucans (fungi), etc. A pattern recognition receptor (PRR) is a metazoan (multicellular) organism's repertoire used to detect and subsequently destroy or otherwise neutralize such invaders. In the simplest terms, single celled PAMPs on the outside are met with PRRs that project outward from metazoan cells or that are free-floating in blood or hemolymph (present often as proteases including complement) and signal to the inside of the cells with the subsequent production of anti-microbial substances or to produce chemical signals (cytokines) to recruit more specialized immune cells to battle the PAMP-containing invaders (including B cells and T cells).

According to Jack and Du Pasquier: "immune receptor repertoires have been selected to 'see' all sorts of different pathogens, and the information they generate can activate powerful effector mechanisms that should certainly be able to destroy all intruders". However, the authors continue, and to paraphrase: the interaction as well as the interface is an "arms race". If one of the pair is modified (e.g. a microbial PAMP), then the other must follow suit (e.g. a metazoan PRR) or risk a decrease in survival.

Many PAMPs are often parts associated with microbial cell wall structures as this is the invariable interface of metazoan and microbe. Endotoxin from the outer cell membrane of Gram negative bacteria has been amongst the most historically recognized PAMP given its broad and systemic effects even as present in minute quantities including cytokine production, fever and coagulation. It can be seen below at right as the major outer barrier in Gram negative bacteria. Each major class of prokaryotes or single-celled organisms including Gram negative bacteria (GNB), Gram positive bacteria (GPB), virions, fungi and protozoans contains a growing list of identified PAMPs. GNB, for example, contain endotoxin, porins, flagella, lipoproteins in the membrane and also include nucleic acids (DNA, RNA) from inside the cell.

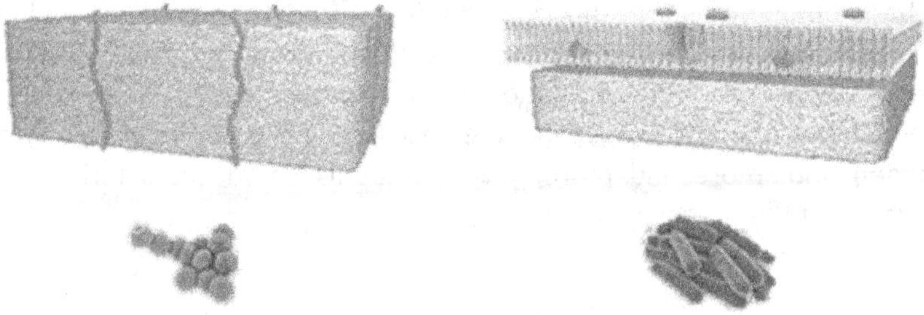

Purple-stained gram-positive (left) and pink-stained gram-negative (right). The outer layer of the dual layered GNB outer membrane is composed of endotoxin (lipopolysaccharide) and phospholipids.
Adapted from http://www.scientificanimations.com/wiki-images/CC BY-SA 4.0. Removed background and added grey scale.

Toll-like receptors (TLRs) "snag" PAMPs in the blood milieu using hydrophobic Leucine rich repeats (LRR), an evolutionarily conserved protein pattern recognition receptor (PRRs). Metazoan innate receptors are available to respond immediately to generic microbial patterns, using Toll-like and complement receptors. Adaptive receptors must be "built" based on initial contact with the enemy. Those B cells (for external antigens) and T cells (for internal cellular antigens) that detect foreign proteins are clonally selected, multiplied and antibodies are secreted and sent out to meet the invaders. This process of production can take a week or more to develop. However, these cells also have memory and serve as the basis for vaccination that may last for decades.

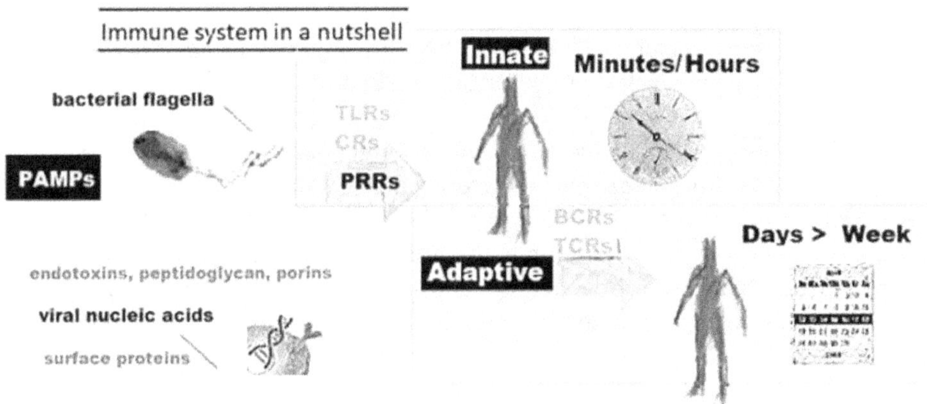

Endotoxin from Gram negative bacteria (GNB) has been the predominate worry of drug manufacturing as a potential contaminant as it is ubiquitous, potent, and highly stable to heat and other forms of destruction. Nanogram and even picogram levels have effects in human blood including cytokine production and coagulation in microvasculature. In drug manufacturing, PAMPs from Gram Positive bacteria (GPB), protists and fungi have not been a great historic concern because they cannot "grow up" quickly in a water-based manufacturing process and do not present either a potent or stable contaminant. Even feared exotoxins from GPB are easily destroyed by heat. However, when manufacturing moves away from its water-based historical roots then other PAMP types can assume a more relevant place of concern including exotoxins. Viral contaminants have been a historical concern for aseptically produced products and biologics but much less so for terminally sterilized products.

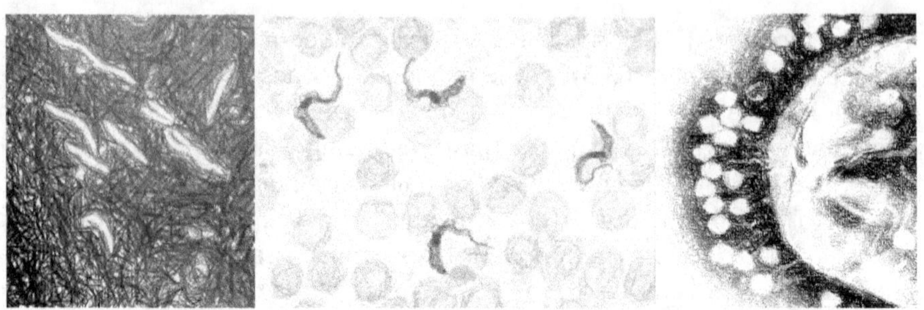

Multicellular and parasitic or saprophitic forms can also present PAMPs to various metazoans. *Borrelia burgdorferi*, the bacteria that causes Lyme disease, is transmitted by *Ixodes* ticks (left). And protozoan, single-celled parasites are also guarded against via PRRs. The protozoan endoparasite Trypanosoma among human red blood cells is shown (center). Bacteriophage attacking a bacterial cell is shown at right (by Dr Graham Beards, CC BY-SA 3.0).

To distinguish PAMPs as pieces of whole (infectious) organisms that must be guarded against (i.e. in drug manufacturing of injectable solutions) via PRRs from a metazoan immune perspective, the following illustrations depict an example molecular artifact that is a porin which are sets of proteins (typically they form as trimers) that allow for the passage of solutes in and out of the microbial outer membrane which otherwise presents a barrier that is very difficult to penetrate.

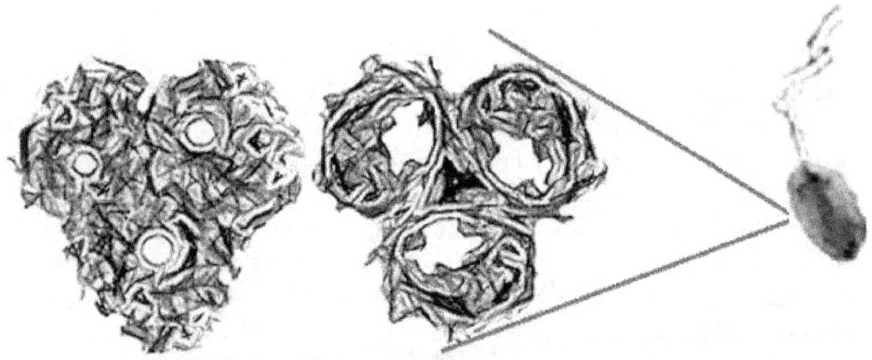

Porins dominate as surface features of many bacteria including GNB. The porin trimer at left above is made up of individual proteins. Right, flagella on a Gram negative bacterial cell. Porin proteins are detected by TLR2 in mammals. Flagella are made of polymers of the flagellin protein which is a PAMP ligand of TLR5. Endotoxin, not shown, is detected by TLR4.

An artistic rendering of an atomic force microphotograph (AFM) is shown below.[9] The spacing and prevalence on the surface of most GNB is a surprising divergence from the typical textbook model of the bacterial surface as shown below at right with few porins. Also the lipopolysaccharide (LPS) molecules should be shown to more thickly cover the surface as in everywhere the porin "holes" are not.

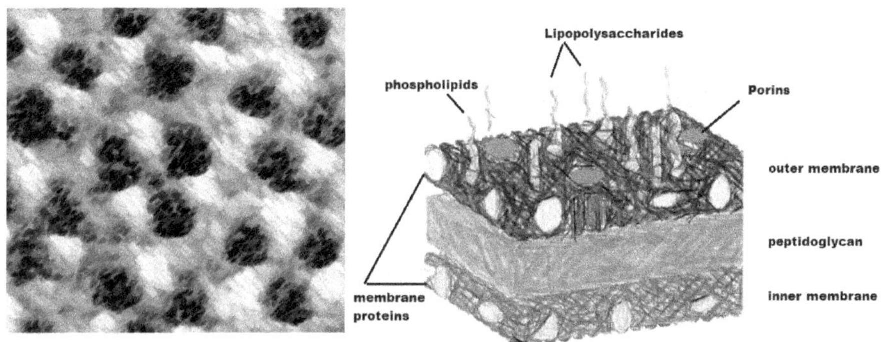

Only a small number of PAMPs have been identified and associated with testing in terms of pharmaceutical microbiological control. As a class, pyrogens (fever causing substances) are the largest class that can be detected, however, the levels of detection, with the exception of endotoxin, are not highly sensitive. As a general rule, in clinical sepsis, the causative agents have been found to be equally Gram positive and Gram negative PAMPs. Additionally, while pyrogenicity has been the solitary concern, now with the advent of biologics immunogenicity concerns have come to predominate. The addition of porins as a quality marker could help ensure the detection of GNB artifacts.[10] The changing expectations of contaminant preclusion can be seen in concerns with protein aggregates and monoclonal antibody paratope sequences.

As further demonstration of the relatedness of single celled and multicellular life, the presence of porins in mitochondria are believed to be derived from the capture of an early single celled organism for use in producing energy for metazoans. The "**endosymbiotic theory**" is described below by Zeth and Thein.

> The endosymbiotic theory forms the basis of our current understanding for the process of how mitochondria and Gram negative bacteria are related by evolution. According to this theory, the mitochondrial organelle originated from a prokaryotic organism and independently developed over a time range of approx. 1.5 billion of years. The overall structure has not changed dramatically and both species have a similar size range of 1– 10 μm with some rare exceptions. The protein class of porins forms the major protein constituents of bacteria…. Although several bacterial genes encoding porin molecules are observed in the genome (e.g. for *Escherichia coli* OmpG, OmpF, PhoE, OmpC and maltoporin, where Omp is outer membrane protein), often one porin variant (e.g. OmpF or maltoporin in *E. coli*) is strongly abundant. In bacteria the number of porin copies was determined to be up to 100,000 per cell wall, whereas in mitochondria the number of VDAC channels was

estimated to be in the range 1000–10,000, depending on the organelle's origin.[11]

Hiller et al.[12] described the VDAC (voltage-dependent anion channel) as: "the most abundant protein of the mitochondrial outer membrane" and VDAC "facilitates the exchange of ions and molecules between mitochondria and cytosol and is regulated by interactions with other proteins and small molecules".

5. What are some of the defenses devised by primitive and complex metazoans as a result of eons of infection?

5.1 Early metazoans

Early metazoans figured out several ways to detect and subsequently destroy invading microorganisms:

(a) the use of **proteases** to chop and degrade proteins of invading organisms and PAMPs.

(b) the use of repeated **hydrophobic peptide sequences**, including Leucine Rich Repeats (LRRs), to bind and snag and subsequently surround and inactivate or destroy foreign substances

(c) the devising and use of **anti-microbial proteins**

(d) **phagocytosis** began as an engulfment of single-celled bacteria by single-celled protists (eukaryotes) but can also be found as a tactic retained in all metazoans many of which have specialized cells to perform this function

(e) **allorecognition** as a primitive form of histocompatibility.

While these metazoan tactics have been thoroughly disseminated throughout the animal kingdom, the first possessors of such tools of destruction could be considered to be bacterial "restriction endonucleases" (RE) which they used to fend off bacteriophage attack by using these enzymes to chop up the viral DNA/RNA. The REs will assume a large role later because they have enabled the advent of recombinant technologies associated with creating biologic drugs.

5.2 Tools used by early metazoans to battle prokaryotes are listed below.

(a) Proteases are used to chop and degrade other proteins

The thing about proteases, which chop protein sequences either indiscriminately or discriminately at specific amino acid sequence combinations (i.e. serine protease), is that they are a "tool" that can be dangerous to the self-cell's own proteins. Their chopping power must be harnessed. This has been achieved throughout the animal kingdom by harboring them in an inactive form and further making their activation very specific to microbial ingress.

(b) Hydrophobic peptide sequences, including Leucine Rich Repeats (LRRs) are used to bind and thus snag and bind foreign substances has culminated in mammalian Toll-like receptors. The origin and development of LRRs and TLRs (including *Drosophila* Toll) will be the topic of more detailed discussion in the next chapter.

The very basic function of hydrophilic and hydrophobic amino acids govern the interactions used by PAMP and PRR alike as contained in both host proteins [including anti-microbial proteins (AMPs) and PRRs like TLRs and also antibody receptors (complementary determining regions or CDRs)] AND PAMPs (peptides, proteins and lipoproteins etc.) that are bound by PRR. A myriad of three dimensional configurations of proteins is achieved automatically by stringing together amino acids as a product of transcription and translation from DNA and mRNA respectively. This "automatic" springing to "life" of proteins when they are strung together is due to the fact that each of the 21 amino acids used to form the protein each have very different properties and begin immediately after formation to exert an effect one another.

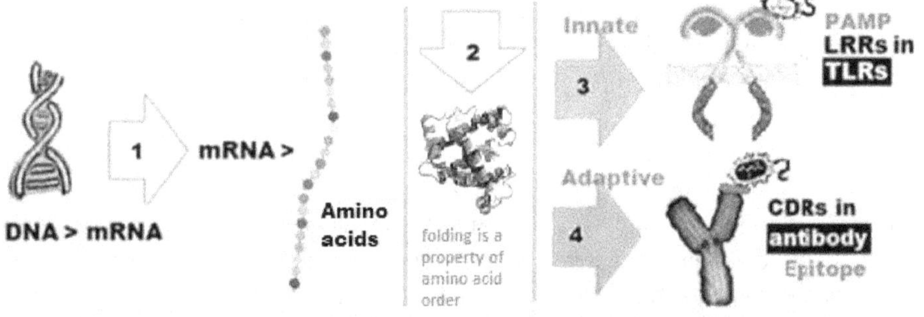

The sum of the electrostatic forces within a protein forms it's "isoelectric point" (pI) which is also function of the pH of solution it is in. There is some thought that the pI/pH dynamic has a role in specific diseases where malformed proteins aggregate and accumulate: "abnormal proteoforms in the blood", including those "such as Alzheimer's disease (AD), Amyotrophic lateral sclerosis, prion disease, Creutzfeldt-Jakob's disease, Parkinson's disease (PD), amyloidosis and a wide range of other disorders".[13] This not-so-simple dynamic, the forces within a protein molecule, form the basis of the PRR and PAMP dynamic in both innate and adaptive workhorse structures (TLRs and Antibody) as show above. The malformation of proteins that are disfigured (relative to the functional form) due

to the substitution of a single (sickle cell anemia[1]) or multiple (Huntington's[2])
amino acid are a major source of various inherited diseases.

(c) Anti-microbial proteins
This item has significant overlap with the previous, hydrophobic protein
sequences, as the mode of action of antimicrobials is often via intercalation with
the bacterial membrane.

> The circulating hemolymph in marine invertebrates contains biologically
> active substances such as complement, lectins, clotting factors, and
> antimicrobial peptides. All of these factors contribute to a self-defense
> system in marine invertebrates against invading microorganisms, which can
> number up to 10^6 bacteria/ml and 10^9 virus/ml of seawater…[14]

Horseshoe crabs are considered *"living fossils"* and have survived from some 400
million years. Thus their immune system has been studied as a "model system"
representative of those from antiquity.

> Tachyplesin is an antimicrobial peptide recently found in the acid extract of
> hemocytes from the Japanese horseshoe crab (*Tachypleus tridentatus*)… In
> our continuing studies on the peptide, we have found an isopeptide,
> tacyplesin II, and also polyphemusins I and II in hemocytes of the American
> horseshoe crab (*Limulus polyphemus*)… all of these peptides inhibited the
> growth of not only Gram-negative and –positive bacteria but also fungi, such
> as *Candida albicans* M9. [15]

According to Tincu and Taylor, "tachyplesin causes a rapid K^+ efflux from
Escherichia coli cells concurrent with a reduced cell viability by permeabilizing
both bacterial and artificial lipid membranes."[16] The shared peptide structure is
given below.

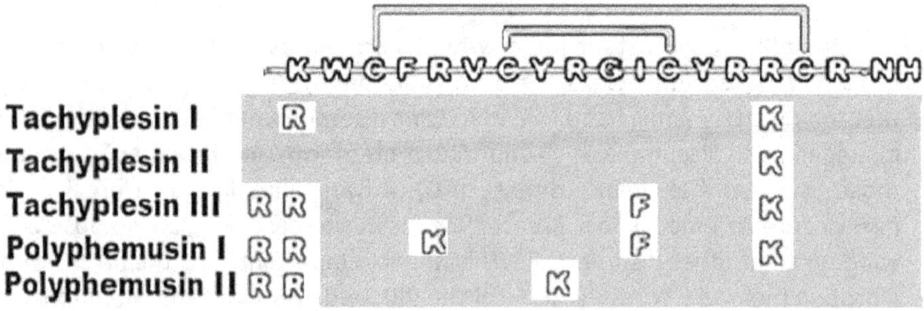

Tachyplesin I		R			K
Tachyplesin II					K
Tachyplesin III	R	R		F	K
Polyphemusin I	R	R	K	F	K
Polyphemusin II	R	R		K	K

------

[1] GAG codon changing to GTG of the β-globin gene
[2] Severity of disease is linked to the number of CAG repeats in the protein

18

Amino acid sequences of tachyplesinns and polyphemusins isolated from horseshoe crabs. Identical amino acids are shown as a solid grey block. The disulfide linkages between Cys-3 and Cys-16 and between Cys-7 and Cys-12 are shown by the heavy double lines connecting the respective cysteine residues. Derived from Tincu and Taylor. Note that K= lysine, R= arginine, W=tryptophan, F= phenylalanine, V= valine, C= cysteine, Y= tyrosine, and G= glycine.

In the structures above, each protein is "bent" by the cysteine amino acids that are attracted to each other in the structure and form a strong disulfide bond. Often proteins with cysteine bonds are secreted proteins. The vast difference in a single amino acid substitution can be seen in that even a single amino acid substitution can bring an extreme change in the overall three dimensional folding of the protein. This is true even for some amino acids that are very close in structure as cysteine has the same structure as serine, but with one of its oxygen atoms replaced by sulfur.

(d) Phagocytosis began as the engulfment of single-celled bacteria by single-celled protists (eukaryotes) but can also be found as a tactic retained in all metazoans, many of which have specialized cells to perform this function. Phagocytosis can be seen in the *Limulus* amebocyte cell (the only hemolymph cell in horseshoe crabs) as well as in *Drosophila* (fruit fly) cells named plasmatocytes or macrophages as well as in mammalian monocytes and macrophage.

Kim et al. describe an anti-microbial peptide in an insect with a wide binding range that serves also as a tag for phagocytosis which is a primitive complement-like activity.

> We characterize a novel pathogen recognition protein obtained from the lepidopteran[3] *Galleria mellonella*. This protein recognizes *Escherichia coli, Micrococcus luteus, and Candida albicans* via specific binding to lipopolysaccharides, lipoteichoic acid, and β-1,3-glucan, respectively. As a multiligand receptor capable of coping with a broad variety of invading pathogens, it is constitutively produced in the fat body, midgut, and integument but not in the hemocytes and is secreted into the hemolymph. The protein was confirmed to be relevant to cellular immune response and to further function as an opsonin that promotes the uptake of invading microorganisms into hemocytes. Our data reveal that the mechanism by which a multiligand receptor recognizes microorganisms contributes substantially to their phagocytosis by hemocytes.[17]

In metazoans, phagocytosis has come to be mediated by complement attachment. As per Jack and Du Pasquier (pg. 61):

---

[3] Butterflies and moths

The neutrophils attracted to a site of infection are there to destroy the pathogen by phagocytosis, and the mechanism of phagocytosis remains basically what it always was. Yet over the course of time it has benefited from numerous technical improvements, and the greatest of these came with the evolution of the complement system. Complement provides a means of labelling particles or molecules that have been identified as "non-self" either by innate immune receptors, such as MBL, or by antibodies.

(e) Allorecognition is the early establishment of a more sophisticated immunity in that it allows a primitive animal, an invertebrate, to distinguish self from not only non-self but from other like-selves. This allorecognition is thought to be the beginnings of "histocompatibility" which will be discussed later on in some detail. A short summary is given by Rosengarten and Nicotra.

> Nearly all colonial marine invertebrates are capable of allorecognition — the ability to distinguish between self and genetically distinct members of the same species. When two or more colonies grow into contact, they either reject each other and compete for the contested space or fuse and form a single, chimeric colony. The specificity of this response is conferred by genetic systems that restrict fusion to self and close kin. Two selective pressures, intraspecific spatial competition between whole colonies and competition between stem cells for access to the germline in fused chimeras, are thought to drive the evolution of extensive polymorphism at invertebrate allorecognition loci. After decades of study, genes controlling allorecognition have been identified in two model systems, the protochordate *Botryllus schlosseri* and the cnidarian *Hydractinia symbiolongicarpus*. In both species, allorecognition specificity is determined by highly polymorphic cell surface molecules...[18]

Another example of allorecognition in metazoans can be seen in *Drosophila* in its on-going battle to resist its larvae from being the victim of parasitic wasps.

> We have been studying the immune response of *Drosophila* against parasitoid wasps as a model system to understand the genomic basis of evolutionary processes. *Drosophila* larvae are host to a variety of parasitoid species that lay an egg in these larvae (Fleury et al., 2009). Once the parasitoid egg hatches (~2–4 days after parasitoid attack, depending on parasitoid species, and temperature), the parasitoid larva starts feeding on the host and kills it. Some species of *Drosophila* have a defense mechanism against parasitoids through an innate immune response, called melanotic encapsulation. This immune response consists of cellular and humoral components that act jointly to sequester and kill the parasitoid egg. Parasitoid attack triggers immune signal transduction pathways that induce (i) the proliferation and differentiation of two classes of hemocytes (i.e.,

insect blood cells) that adhere to the parasitoid egg and to each other, and (ii) the deposition of melanin on the parasitoid egg and the cellular capsule around the parasitoid egg (Lemaitre and Hoffmann, 2007). The host has to complete the full encapsulation and melanization before the parasitoid egg hatches to survive the parasitoid infestation.[19]

5.3 *For more complex metazoans like mammals…*
Inflammation, fever, and coagulation are all affected or initiated by cytokine production as a result of the interaction of PAMP with PRR, and are the hallmarks of infection.   Allorecognition in jawed vertebrates has expanded to a sophisticated system of verifying the "self, non-self" status of all nucleated cells in the body via T-cell, MHC I complexing that will be described in Chapter 3.

Infection or perceived infection (from non-living microbial artifacts as an injection contaminant or sepsis holdover) brings about systemic or whole body reactions that have been used for ages to gauge the severity of infection. One could argue that observations of both fever (i.e. rabbit pyrogen test) as well as coagulation (i.e. coagulation via the *Limulus* test) derive from a pre-molecular understanding of the immune system since they were both developed prior to the discovery of PPRs or elaboration of lymphocyte receptors. A modern understanding of both innate and adaptive immunity didn't begin to arise until the 1990's although both innate and adaptive therapies have been used empirically for decades (empirical is "whatever works") and included aspirin for fever and sera from purposefully infected animals as a treatment for diphtheria (for example).

> TLRs are prototype pattern-recognition receptors (PRRs) that recognize pathogen-associated molecular patterns (PAMPs) from microorganisms or danger-associated molecular patterns (DAMPs) from damaged tissue. The idea that such innate immune receptors exist dates back to 1989, when Charles Janeway predicted, in an important monograph, that PRRs recognizing microbial products link innate and adaptive immunity.[20]

Fever occurs with viral and microbial infections and is the classical sign of infection, whereas coagulation is a later stage complication associated with sepsis. It is first an issue in the microvasculature that can become clogged thus cutting off oxygen to some tissue which, if it persists, can lead to the need to amputate limbs that have experienced oxygen deprivation as shown below. This "walling off" of parts of the organism is reminiscent of the horseshoe crab's response which is to coagulate at the sites of microbial ingress.

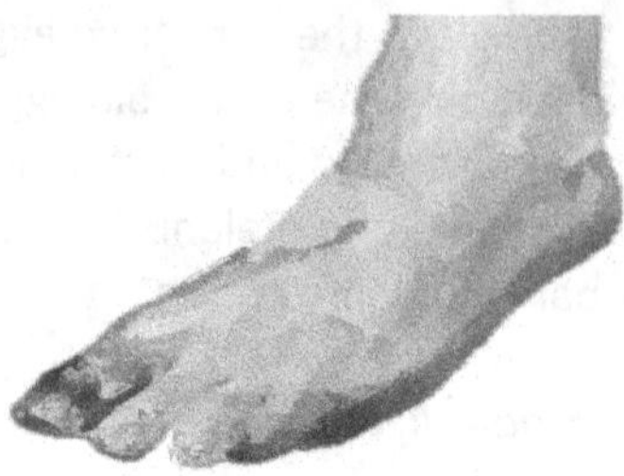

A common result of sepsis is amputation due to the blocking of blood supply in extremity microvasculature.

6.   *What is the timeline of the discovery of pattern recognition receptors (PRRs)?*

Many studies have fed into the modern understanding of PAMPs and PRRs (as the ancient to modern microbial, metazoan interface), however, arguably, six specific events serve as foundational for the current understanding (listed as somewhat overlapping rather than strictly chronological):

(i)   Janeway describes the very idea of the pattern recognition receptor and PAMP interaction in his seminal talk "Approaching the Asymptope" (1989).[24]

(ii) The discovery of *Drosophila* Toll (*dToll*) in establishing the embryotic dorsoventral axis

(iii) The discovery that *dToll* also has an immune function in the adult fruit fly

(iv) The discovery of a human homologue to *dToll* that was first called *hToll*

(v) The discovery that the gene that encodes the *hToll* identified by Janeway et al. in fact produces the protein product (TLR4 receptor) that detects LPS (endotoxin).

(vi) The discovery of multiple *hToll* receptors called "Toll-like receptors" in mammals.

i.   Janeway's 1989 Cold Spring Harbor talk was very influential in kicking off a new view of immunology. Here he outlined his suspicions that static inmate receptors must exist in contrast to variable adaptive receptors that had already been discovered as lymphocyte antibodies that detect antigen. He reasoned that adaptive variable receptors could not be completely trustworthy given that they are the product of random rearrangements made to anticipate a wide variety of possible antigen structures. He called this confirmation the "second signal" that answers the question from adaptive immunity : "can you confirm this is friend or foe?":

The Nonclonal Origins of the Immune System

If effector mechanisms used by lymphocytes in contemporary vertebrate immune systems derived from primitive immune systems lacking the rearranging receptor gene families that allow for global selection, how was effector function regulated in primitive organisms? The most likely possibility is that primitive effector cells bear receptors that allow recognition of certain pathogen-associated molecular patterns that are not found in the host. I term these receptors pattern recognition receptors. In this section, I argue that pattern recognition receptors are still an important part of vertebrate immune systems. However, because they are not clonally distributed on lymphocytes they do not confer immunological memory, the hallmark of the specific immune response. Possible examples of pattern recognition receptors would be the uncharacterized receptors that allow natural killer cells to discriminate between target cells, the antigen-presenting cell receptors acted upon by adjuvants to induce second signals and the many T cell surface molecules of unknown specificity that can induce T-cell activation upon cross-linking (Table 2). I propose that these pattern recognition systems activated effector functions of primitive immune systems prior to the development of rearranging gene families and continue to play a roll in defense today.

Janeway's Table 2. T-cell Surface Molecules That Can Cause Activation upon Cross-linking by Antibody. CD type markers will be discussed in Chapter 4.

T-cell receptor ($\alpha : \beta$ and $\gamma : \delta$)

Thy-1
CD2
CD23
CD27
CD28
CD43
CD44
CD45
CD54
CD73
Ly6

ii. Christiane Nüsslein-Volhard, 1985 proclaimed at the sight of a deformed *Drosophila* larvae, *"Das war ja toll!"* or *"That is crazy (or strange)!"* in German. Her studies were instrumental in demonstrating that Toll had a role in establishing the dorsoventral axis of the developing fly embryo. Larvae segmentation and the low number of chromosomes (4 pairs[4]) allowed the location of many developmental genes including *hedgehog*, *gurken* (German: "cucumbers"), and *Krüppel* ("cripple"), named as they were for the larval appearance when the associated gene was knocked out. Thus Toll was known to exist according to the work done by Nüsslein-Volhard and Wieschaus but there was no suspicion yet that it include an immune related function.

iii. Lemaitre et al. in 1996 published their landmark paper[21] which described the immune function of *Drosophila* and more specifically an immune-related role for Toll that Nüsslein-Volhard and Wieschaus previously identified. Among the most important aspects of the findings was that in adult flies with mutations in dToll, the flies succumbed to infection with the fungus *Aspergillus fumigatus* but not *Escherichia coli*[22] as shown below.

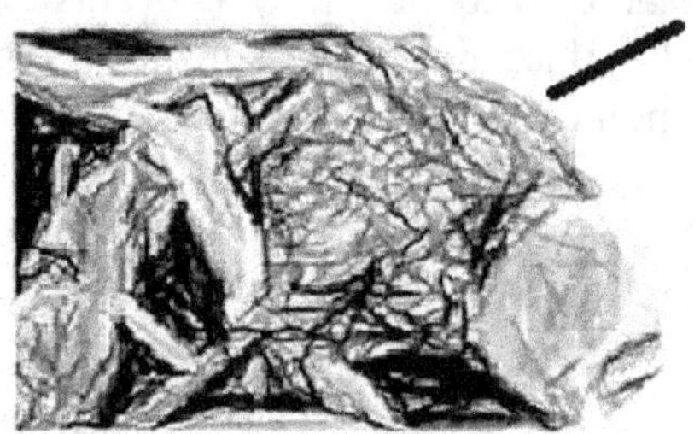

The Hoffman lab's discovery of the *Drosophila* Toll pathway and fungal overgrowth that occurs when Toll is disabled is shown above (mold "fur coat"). Derived from the famous scanning electron-micrograph of a *Drosophila* adult that succumbed to infection by *Aspergillis fumigatus* and is covered with germinating hyphae.

iv. Janeway, Medzhitov, and Preston-Hurlburt, in 1997 compared a gene sequence from the transmembrane part of the *dToll* receptor (TIR) against known sequences in a human genetic database and found a match that they referred to as hToll.

> It thus appears that the immune-response system mediated by Toll represents an ancient host defense mechanism. To investigate the possibility that this pathway has been retained in the immune system of vertebrates, we used sequence and pattern searches of the expressed-sequence tag (EST) database at the National Center for Biotechnology Information (NCBI). A search with a sequence profile of the Toll/IL-1R signaling domain identified a matching sequence in the EST database derived from human fetal

---

[4] Note that *Drosophila* still has 15,500 genes compared to 22,000 for humans, so the density of genes in the chromosomes is high.

liver/spleen library (Genebank accession number H48602, corresponding to clone 202057 from the IMAGE consortium).

Thus, like *dToll, hToll* is a type I transmembrane protein with an extracellular motif consisting of a leucine-rich repeat (LRR) domain, and a cytoplasmic domain homologous to the cytoplasmic domain of the human interleukin (IL)-1 receptor.

v. Beutler et al. in 1998 used a specific method: "positional cloning work involving the C3H/HeJ and C57BL/10ScCr mouse strains" to determine the gene encoding the receptor responsible for detecting LPS.[23]

> Mutations of the gene Lps selectively impede lipopolysaccharide (LPS) signal transduction in C3H/HeJ and C57BL/10ScCr mice, rendering them resistant to endotoxin yet highly susceptible to Gram-negative infection. The codominant Lps^d allele of C3H/HeJ mice was shown to correspond to a missense mutation in the third exon of the Toll-like receptor-4 gene (Tlr4), predicted to replace proline with histidine at position 712 of the polypeptide chain. C57BL/10ScCr mice are homozygous for a null mutation of Tlr4. Thus, the mammalian Tlr4 protein has been adapted primarily to subserve the recognition of LPS and presumably transduces the LPS signal across the plasma membrane. Destructive mutations of Tlr4 predispose to the development of Gram-negative sepsis, leaving most aspects of immune function intact.

And…

> The solitary nature of the LPS signal transduction pathway was indicated from the very start by the fact that mutations at a single locus (Lps) are able to completely abolish LPS signal transduction, as witnessed in C3H/HeJ and C57BL/10ScCr mice. It is now known that in the latter strain, the Tlr4 locus is deleted entirely, while in the former strain, a point mutation modifies the cytoplasmic domain of the Tlr4 receptor so as to yield co-dominant inhibition of LPS signaling in heterozygotes. Insofar as the modification of a single amino acid is sufficient to wreak such a profound effect on LPS signaling, it may be said that accessory pathways are of negligible importance.

So even though Charles Janeway proposed that PRRs likely existed in his 1989 famous podium presentation with subsequent paper (*Approaching the Asymptote?*[24]), it wasn't until 1997 that the existence of PRRs were proven. *This is such a short time ago!* It turns out that what they found (*hToll*) was indeed the LPS receptor, TLR4. Therefore, the PRR story is very much also a story about endotoxin.

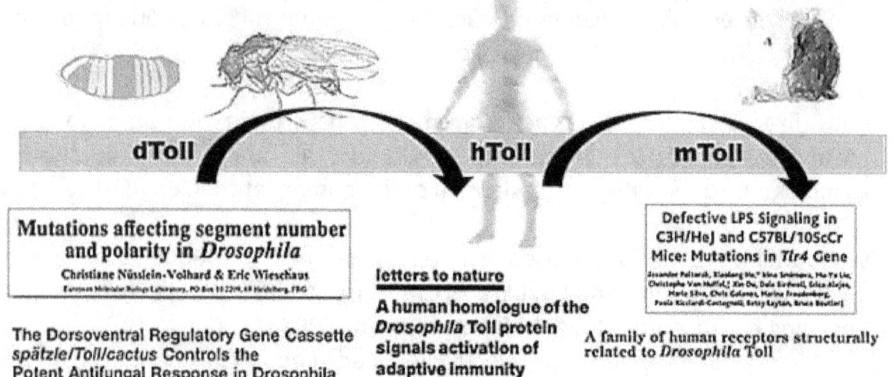

Milestone papers on the road to establishing and understanding innate immunity.

It is apparent that the detection cascade that includes *dToll* is not all that different from the horseshoe crab's mechanism of detecting endotoxin via the Factor C cascade. Thus, the evolutionarily conserved proteins that served *Limulus*, then *Drosophila*, and now mammals have all continued to evolve yet the most modern forms retain some relationship to those from antiquity. The use of the same protein (*dToll*) for two diverse functions (dorsoventral axis formation and immunity) in *Drosophila* shows the extreme economy of the evolutionary process. Another aspect of this process that has been instrumental in furthering the understanding of genetic body development can be seen in the modularity of (arthropod) structures where a leg can be grown in place of an antenna on the *Drosophila* head by simply substituting one gene for another along the linear axis of embryo formation (as shown below).

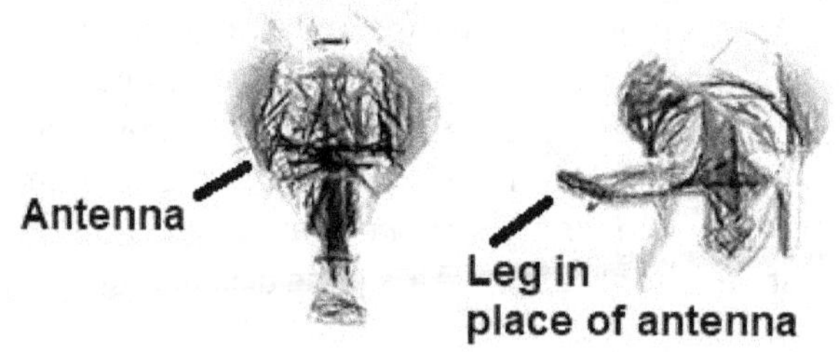

The Nobel Prize in Physiology or Medicine 1995 was awarded to Edward B. Lewis, Christiane Nüsslein-Volhard and Eric F. Wieschaus "for their discoveries concerning the genetic control of early embryonic development". This demonstrated the Hox gene control over body patterning and, as per the pictures above, the modularity of appendages where Lewis' lab was able to transfer a leg in place of an antenna. Derived from:

http://learn.genetics.utah.edu/content/basics/hoxgenes/.[5,6]

vi. The discovery of multiple TLRs by Rock et al. in 1998 established the current paradigm of multiple defense PRRs with multiple ligands bound. An estimate has been given of around 100 PAMPs bound per ten (in man) TLRs gives around 1,000 different PAMPs recognized via the Toll-like receptor part of the inmate immune system alone. Remember, TLRs bind highly evolutionarily conserved structures on, typically, prokaryotes whereas hypervariable structures such as surface structures on flu viruses necessitate the use of hypervariable lymphocyte receptors (a product of which include antibodies from BCRs).

The discovery of the immune portion of this series of events occurred in a short period of time (1996–1998) and the outcomes were as predicted in Janeway's 1989 symposium lecture which included the following propositions (a-c):

(a) Metazoans possess "pattern recognition receptors" (PRR) to detect conserved microbial structures. LPS is the prototypical PAMP as detected by PRRs. Of the ten TLRs in man, not a single PRR was known at the time of Janeway's famous lecture. Since these PAMP structures are integral to the organism that possess it, they are evolutionarily conserved. In this way, the effort expended to develop PRR against such PAMPs is an investment and thus this investment is made into "sure thing" type structures.

(b) PAMPs exist as the molecularly conserved markers for prokaryotic invasion (includes LPS, peptidoglycan, nucleic acids, flagellin, etc.) and are detected by PRRs. Endotoxin was well known by its fever and *Limulus* clotting ability without a corresponding knowledge of the molecular mechanism (TLR4) needed to "see" it in mammals.

(c) Janeway also predicted the control of the adaptive immune system via the innate immune system at a time when most viewed them as largely non-overlapping. Lymphocytes require the activation of costimulatory PRRs of the innate immune system. This last prediction of Janeway yet to be described here will be elaborated in Chapters 3 and 4. The basic interplay of microbial PAMPs with metazoan PRRs continues to be fundamental to the ongoing discussion.

---

[5] This is an excellent quick overview of Homeotic Genes and Body Patterns.

[6] See also: Modeling congenital disease and inborn errors of development in *Drosophila melanogaster,* Matthew J. Moulton and Anthea Letsou, here:
https://dmm.biologists.org/content/dmm/9/3/253.full.pdf

7. How do we <u>know</u> that modern immune systems are made from modified, even cobbled together, ancient immune structures?

7.1 Quick proof

This paradigm is generally accepted, but if a quick "proof" is desired as a reminder, then a comparative look at *Limulus, Drosophila* versus mammals gives the necessary background.

Dobzhansky made the famous observation that *"Nothing in biology makes sense except in the light of evolution"*.[25] This is also true of the immune system as viewed in an ancient versus modern context. Structures in complex systems are derived from structures that worked in simpler systems in some previous form and even as we have seen as derived from other functions (*Drosophila* dual use of Toll previously discussed). While not directly descended, mammals and horseshoe crabs share a distant common ancestry that extends very far back and predates the protostome and deuterostome split some 558 million years ago. Protostomes and deuterostomes are split from the bilatria right before the Cambrian period (541-585 mya).

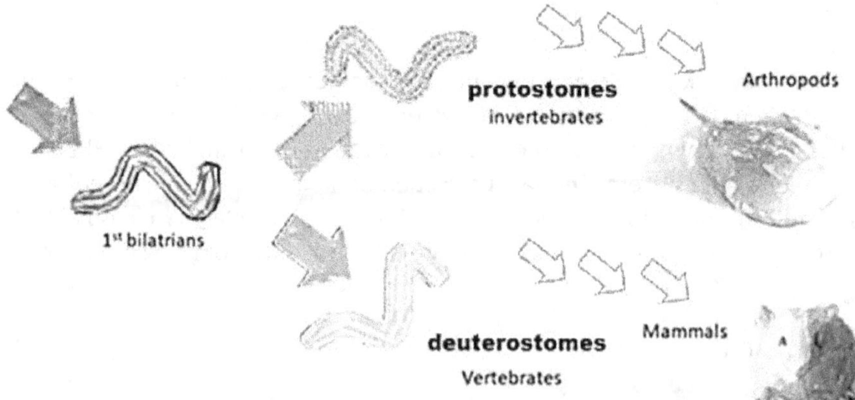

In Greek **protostome** means *"first mouth"* whereas **deuterostome** means "*second mouth*". The major difference between deuterostomes and protostomes are derived from their embryonic development in that the embryological origins of the mouth are formed first in protostomes and secondly (after the anus) in deuterostomes (with a few exceptions).

How can ancient animals be used to model immune evolution? Given the eons of divergence between horseshoe crabs and mammals, it is not obvious why there should be *any* overlap in structure or function other than sharing the somewhat advanced yet primitive pre-divide basic metabolic system that existed as the precursor for evolution in either direction. The "anthropomorphic" tags of "advanced" and "primitive" are often objected to, yet, how else can one understand the great gain in utility. This gain of utility is enabled by the gain in useful evolutionary complexity which is the successful perception and interaction with reality (the physical and invisible universe).

The horseshoe crab has a primitive blood/hemolymph system that shares some common features with mammalian blood. A single circulating blood cell is able to phagocytize bacteria and de-granulate into the hemolymph serine protease clotting factors that are triggered by the presence of lipopolysaccharide.

Consider that *Drosophila* and humans share 75% of disease-related genes.
The fact that the ancestor of vertebrate and invertebrate model organisms was a highly evolved creature which had already invented complex interacting systems controlling development, physiology, and behavior has profound implications for medical genetics. The central points that we explore in this chapter can be broadly put into two categories: (1) the great advantages of model organisms for identifying and understanding genes that are altered in heritable human diseases and (2) the functions of many of those genes and the evidence that they were present in the ancestral bilateral organisms and have remained largely intact in both vertebrate and invertebrate lineages during the ensuing course of evolution. In the course of discussing these points, we review the compelling evidence that developmentally important genes have been phylogenetically conserved and the likelihood that developmental disorders in humans will often involve genes controlling similar morphogenetic processes in vertebrates and invertebrates. A systematic analysis of human disease gene homologs in Drosophila supports this view since 75% of human disease genes are structurally related to genes present in Drosophila and more than a third of these human genes are highly related to their fruit fly counterparts (Bernards and Hariharan, 2001; Reiter et al., 2001; Chien et al., 2002).[26]

Where can one view the remnants of commonality between primitive and advanced metazoans from an evolutionary vantage? As an example, *Limulus* and mammalian proteins have been compared and the commonality (called homology) between proteins (ancient to modern) are shown below as derived from Muta et al.[27] Many of the blood factor and blood complement proteins show significant sequence homology.

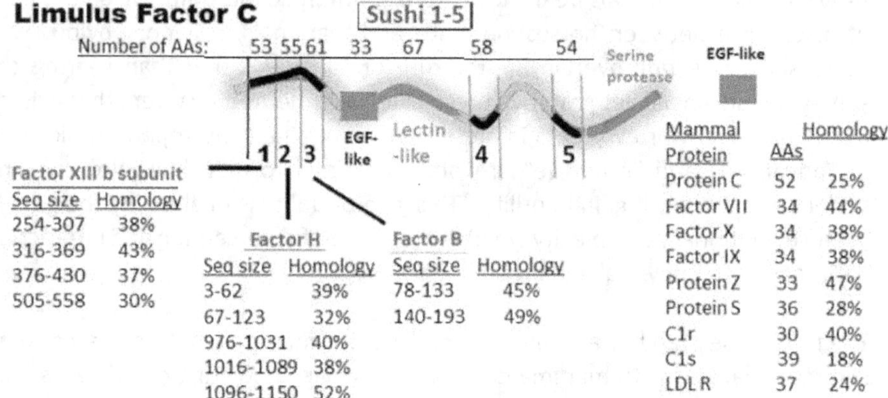

Factor XIII b subunit

Seq size	Homology
254-307	38%
316-369	43%
376-430	37%
505-558	30%

Factor H

Seq size	Homology
3-62	39%
67-123	32%
976-1031	40%
1016-1089	38%
1096-1150	52%

Factor B

Seq size	Homology
78-133	45%
140-193	49%

Mammal Protein	AAs	Homology
Protein C	52	25%
Factor VII	34	44%
Factor X	34	38%
Factor IX	34	38%
Protein Z	33	47%
Protein S	36	28%
C1r	30	40%
C1s	39	18%
LDL R	37	24%

EGF is epidermal growth factor. AA is amino acid. Interpretation example: Factor B is a component of the alternative cascade of the human complement system. Factor B consists of a single chain of 764 amino acids and according to the chart sequences 78 to 133 (55 AAs) and 140 to 193 (53 AAs) when lined up with Sushi 3 domain show 45% and 49% homology (same amino acids) as those in Factor C (*Limulus*) [as compared to human blood protein Factor B (Human)]. This is obviously more than coincidence or even convergent evolution but rather remnants of an ancient precursor protein.

From the cartoon below one can see that the mammalian Factor B molecule in blood also shares the 3X sushi domain structure (CCP is synonymous for Sushi) with *Limulus* factor C. VWFA is von Willebrand factor type A and SP is serine protease. Derived from Le et al.[28]

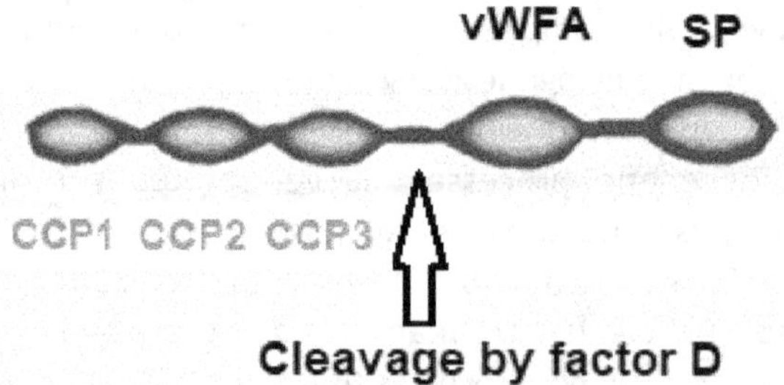

In terms of the "how?" of evolving immune architecture, we can see the generally accepted process below where the hypothetical **urbilaterian** (German **ur** is 'original') is the last common ancestor of all animals having a bilateral symmetry (the body has a mirrored image).

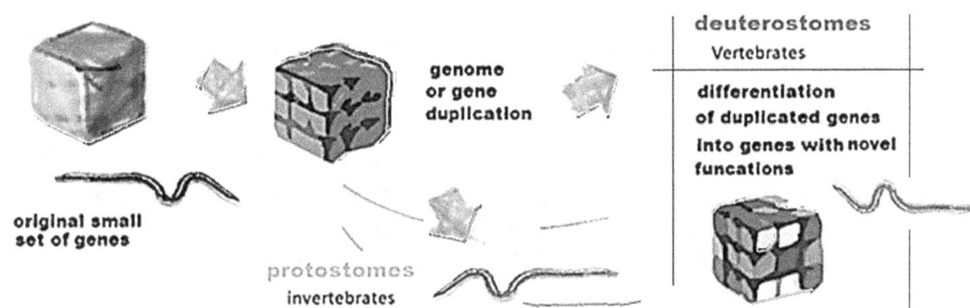

Genome duplication with subsequent mutation where non-essential (duplicate genes) can (be allowed to) change to acquire new functions, and is a generally accepted means by which evolution makes big leaps of change.

7.2 Terrestrialization

Another way to "see" evolution at work includes the coming of age of animals that went from sea to land, known as terrestrialization. According to Sharma: [29]

> Terrestrialization is one of the most iconic and evocative phenomena in evolutionary biology for scientists and laypersons alike. The conquest of land and its concomitant expansion of ecological niche space are associated with a suite of requisite adaptations for such biological processes as locomotion, reproduction, and aerial respiration. While partial or completely terrestrial life histories have evolved several times in Metazoa (e.g., vertebrates, arthropods, annelids, mollusks), the phylum Arthropoda is exceptional in this regard, with at least seven major terrestrialization events occurring at various time scales and phylogenetic depths since the Paleozoic. The impact of terrestrialization on arthropod diversification is considerable, with terrestrial species outnumbering aquatic and marine counterparts by a factor of nearly 17 to 1.

The predominance of terrestrial arthropods is conveyed by the figure below derived from Sharma.

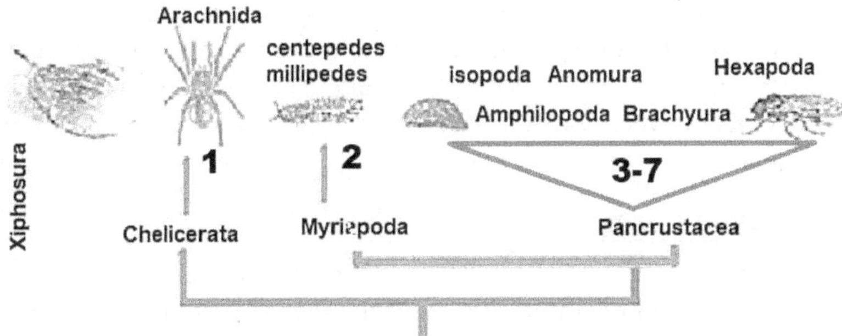

The seven numerically designated events are thought to represent seven different terrestrialization events for the specific arthropods shown which pre-date the animals (from fish) but post-date plants by far.

The obvious segmentation and thus modularization of arthropods has allowed, evolutionarily, for the abundant forms that now exist by the shuffling of hox genes according to the compartmentalization of proteins and other gene products within segments. This is shown below as based upon the figure by Damen et al.[30]

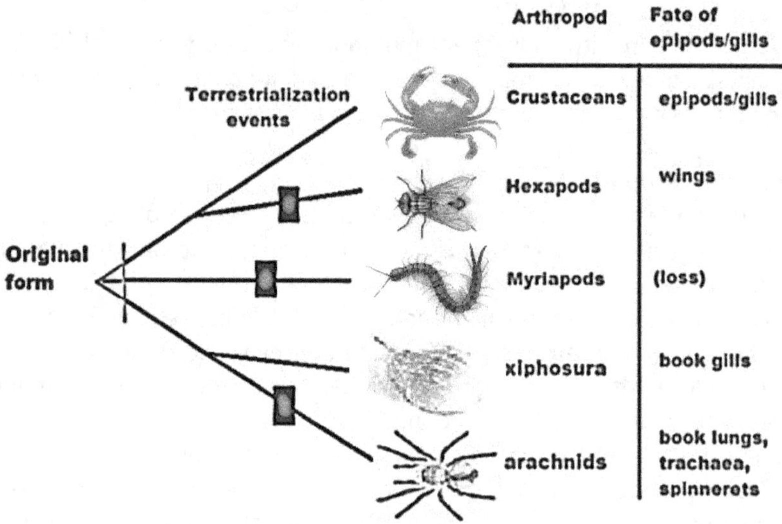

7.3 Models of Infection

There are significant similarities between the horseshoe crab hemolymph and mammalian blood. Of course, there are many dissimilarities. But in terms of using a simpler animal (a single circulating cell versus a dozen in mammals) to model (as a verb) complex actions that occur in higher animals, the horseshoe crab has been an important tool for the modeling of the coagulation of its "blood". Consider that if one only had mammalian blood to study, then certain wrong assumptions might predominate. Namely, the idea that oxygen carry in red blood cells (iron in hemoglobin) and immune cell function in white blood cells represents a modern occurrence as in *Limulus* a single cell performs only innate immune functions and oxygen is carried via copper in the hemolymph separate from cellular attachment. This copper carry gives the horseshoe crab's blood its baby blue color.

Mammals, when wounded or cut, have platelets that are attracted to the wound. They have thrombin receptors that bind thrombin (a serine protease) which is present in the serum. Fibrin is converted by thrombin to long fibers or strands of fibrin that form a clot as bound to the platelets. The horseshoe crab has a simple serine protease cascade that, when triggered by a wound or bacterial endotoxin ingress, the blood quickly clots around the wound or the bacterial ingress. This is the mechanism from which the *in vitro* lab test (LAL) has come.[31]

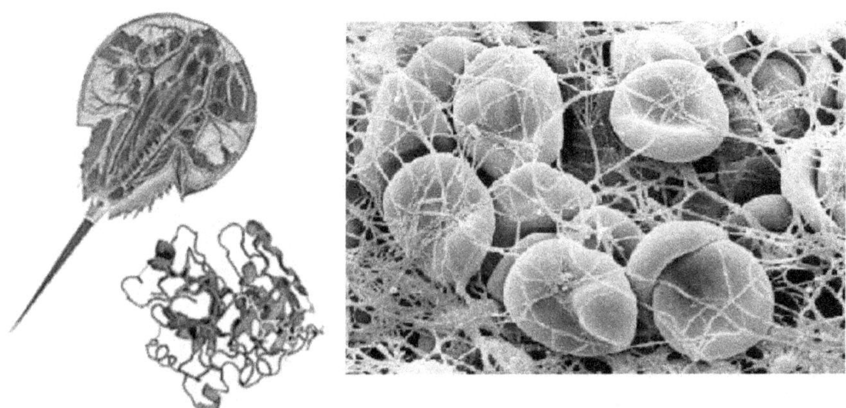

Limulus (left) blue blood (hemolymph) clots when exposed to minute amounts of endotoxin. Similarly, mammalian blood clots in the microvasculature in sepsis (DIC, disseminated intravascular coagulation) patients. *Right,* Scanning electron microscope image of blood clot. The clothing as shown above occurs during microbial infection and can lead to clogging of the microvasculature as in sepsis where extremities including toes and fingers or arms and legs or nose must be amputated.

See simplified mammalian coagulation pathway (below left) and horseshoe crab gelation cascade (below right). According to Muta et al. *Limulus* Factor C is ~37% homologous to mammalian <u>thrombin</u> (central location in left figure).

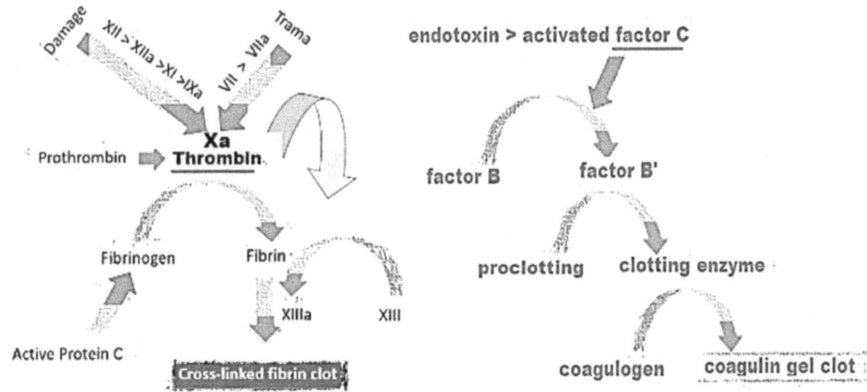

7.4 *Immune system evolution*

While there are no "fossils" to show immune system evolution, in effect the systems that remain today (extant) that are primitive give some genetic "breadcrumbs" that can infer the change that has occurred from some eons ago. Wertheim gives a good idea how difficult this change may be, given the number of interconnected systems involved. For every receptor there is a "signal transduction cascade", a set of "transcription factors" as well as "effector molecules" and/or "specialized cells", as shown below.

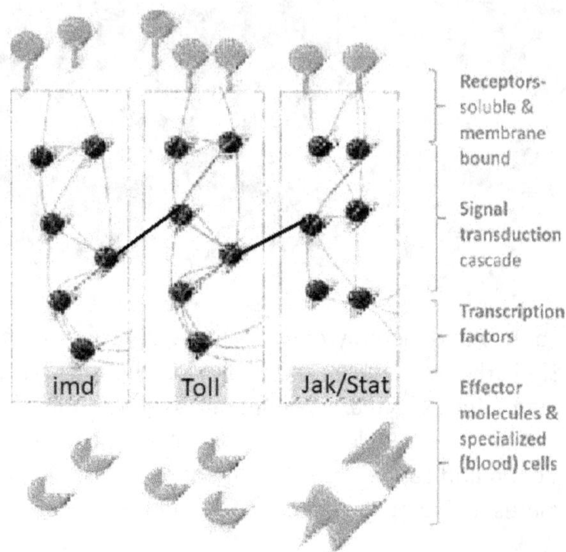

Derived from Wertheim (A only). Schematic representation of the genetic networks in immunity. Several interconnected networks coordinate the responses to an immune challenge. These networks consist of proteins that interact with each in a signal transduction cascade to regulate the expression of transcription factors. The activation of the core signal transduction pathways (e.g., IMD, Toll, or Jak/Stat as labeled) results in the production of effector molecules, such as antimicrobial peptides (represented by pie-shaped symbols) and the proliferation and differentiation of specialized (blood) cells (cloud-shaped figures). Extracellular and membrane-bound receptor molecules (top) induce the pathways. The activity can be further modulated by many other proteins that interact with the pathways and cross-talk with other pathways and genetic networks (indicated by the lines among proteins).

Chapter 2. Mammalian innate immunity

8. *How does the body "see" infectious agents as well as microbially derived non-infectious agents (PAMPs)?*

Having come some ways now from ancient metazoan origins we can focus on mammalian innate immunity here and adaptive immunity in the next Chapter.

Today it is known that there are 12 TLRs in mammals and 10 specifically in man that detect at least a thousand known PAMPs as well as DAMPs (damage-associated molecular patterns). As we will see in the next chapter, though detecting a thousand PAMPs is a great start, especially for short lived animals, it is not enough to ensure long-lived tetrapods survival from prokaryotic onslaught. For this the antibody response is also needed.

The key to metazoan targeting of prokaryotic structures is to pick the best structures, the most conserved and generalized structures that prokaryotes need to survive. These basic structures include endotoxin, exotoxins, various cell wall materials, flagella, beta-glucans, nucleic acids, etc. Given the high rate of reproduction (every 20 minutes in some cases) of prokaryotes and thus the high probability of mutation of microbial structures, then it is critical for metazoan immune systems to target those structures which are highly enduring, at least from the innate immune vantage.

The TLR architecture has been determined to exist in both homodimers and heterodimers where TLR4 is an example of a homodimer that uses a dimer made of mirror units whereas the dimers formed from different TLR structures are not mirrored as shown below by assigned numbers.

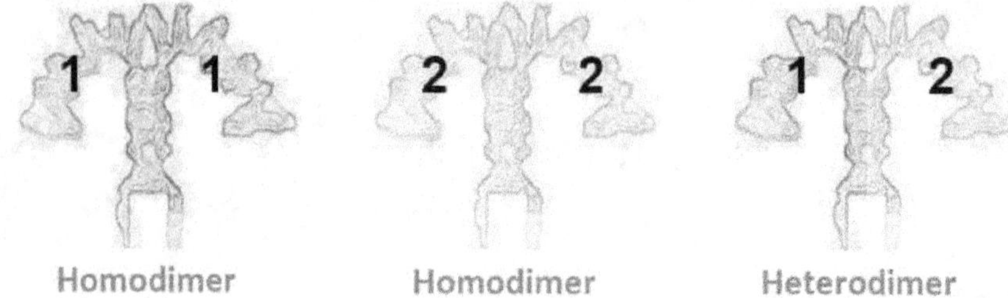

Some homodimers (TLR1 left and TLR2 center) can combine to form heterodimers (right) capable of detecting PAMPs that are a different set than those detected by the respective homodimers. TLR4 is a homodimer, though there are a couple of reports of it being used as a heterodimer.

A very partial list of PAMPs associated with various classes of prokaryotes are listed below.

Viral	Fungal	Gram Positive	Gram Negative
		LTA	
Proteins	Zymosan	Peptidoglycan	Flagella
ssRNA	Mannan	Bacterial proteins / Glycoproteins	Membrane vesicles
dsRNA		Exotoxins	Endotoxins
ssDNA		Porins	Porins
dsDNA		dsDNA (unmethylated CpG-rich)	

The mechanisms of ligand attachment vary greatly amongst the various TLRs as can be seen below. Some attach horizontally across both LRR branches, some attach to accessories molecules such as LPS (triangle) inside MD-2 (black ball) which attaches to TLR4 in pairs. Others have very small differences that cause them to bind an entirely different TLR as seen in Pam3CSK4 binding to TLR2/TLR1 versus Pam2CSK4 binding to TLR2/TLR6.

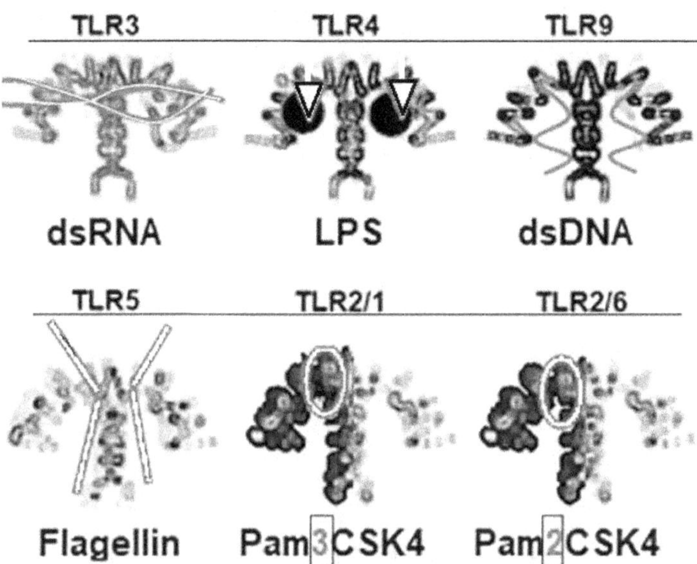

Given that different tissues and immune cell types have different roles and placement in the body and given that the TLRs are genetically modular, then it is not surprising that different receptors will occur on different cells. This allows for a partitioning effect where, for example, tissues that contact the open environment (intestines, esophagus, lungs) are expected to contain a fuller repertoire than other tissues. It's apparent that not only has one or a couple of initial TLRs become a host of TLRs that bind a wide variety of PAMPs, it is also clear that some functional ones have also been lost, such as TLR11 in man.[32] Thus, even as we study, evolution continues to improvise new solutions and suffer new setbacks as well.

Each TLR as it interfaces PAMPs is made up of so-called Leucine-rich repeats (LRRs) which are conserved across the plant and animal kingdoms as building blocks of various different PRRs.

The leucine-rich repeats (LRR)- containing domain is evolutionarily conserved in many proteins associated with innate immunity in plants, invertebrates and vertebrates. Serving as a first line of defense, the innate immune response is initiated through the sensing of pathogen-associated molecular patterns (PAMPs). In plants, NBS (nucleotide-binding site)-LRR proteins provide recognition of pathogen products of avirulence (AVR) genes. LRRs also promote interaction between LRR proteins as observed in receptor-coreceptor complexes. In mammals, toll-like receptors (TLRs) and NOD-like receptors (NLRs) through their LRR domain, sense molecular determinants from a structurally diverse set of bacterial, fungal, parasite and viral-derived components. In humans, at least 34 LRR proteins are implicated in diseases. Most LRR domains consist of 2–45 leucine-rich repeats, with each repeat about 20–30 residues long. Structurally, LRR domains adopt an arc or horseshoe shape, with the concave face consisting of parallel β-strands and the convex face representing a more variable region of secondary structures including helices. Apart from the TLRs and NLRs, most of the 375 human LRR proteins remain uncharacterized functionally.[33]

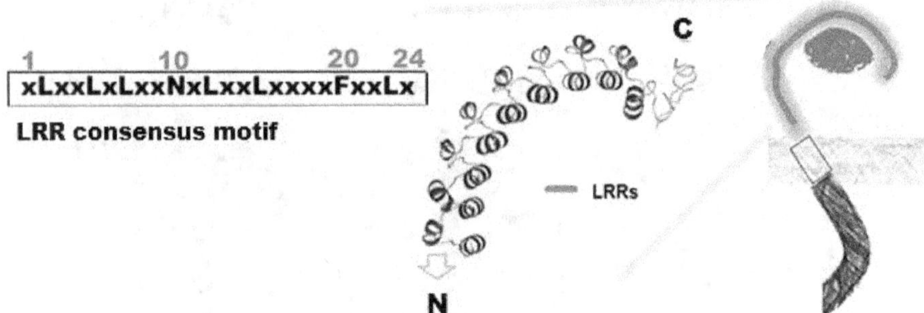

The ancient and basic structure of the LRR at left is shown as a critical structural domain of the modern TLR. LRR consensus motif sequence as given by Robinson and Moehle.[34]   X is any amino acid. L = Leucine. N = Asparagine. F = Phenylalanine.

9. What is the molecular interaction between the external "symbol" (PAMP) and the internal interpretation of insult via PRR?

Both **TLRs** and **complement** are good examples of metazoan defense molecules interfacing PAMPs directly and effectively. In addition to TLR's which cover a broad swath of the microbial PAMP world, complement in the human blood system serves a basic rule: "...everything that is not specifically protected has to be attacked" as per Merle.

> Complement cascade is activated immediately after encountering the pathogen. Hence, complement participates in pathogen opsonization, tagging it for engulfment by antigen presenting cells (APC); it plays a central role in the inflammatory process and modulates the activity of T-and B-cells. After generation of pathogen specific antibodies, complement contributes in the clearance of immune complexes and pathogen elimination. Studies over the years demonstrated that complement takes part in nearly every step of the immune reaction and that it deserves a central position in the immunological research. Unfortunately, the lack of coherence in complement proteins nomenclature and the complexity of the enzymatic cascade render complement one of the "most complicated and incomprehensible" parts of immunology and is frequently avoided by students and scientists...

The main complement rule is that "everything that is not specifically protected has to be attacked". Host cells carry an armamentarium of "don't attack me" molecules, which are either expressed by the cell or recruited to the cell membrane from the plasma. Therefore, any cell, debris, microorganism, or artificial material lacking these molecules (and carrying –OH or –NH2 chemical groups, which is the case of all biological and most synthetic materials) will represent an "activating surface" and will support complement deposition, i.e., covalent binding of C3b, an activated form of the central complement component C3.[35]

This interface of complement with microbe as existing in mammalian complement has also been found fairly recently to exist in primitive form and function in *Carcinoscorpius* (one of four extant horseshoe crabs including *Limulus*). According to Zhu et al.

> The complement system has been thought to originate exclusively in the deuterostomes. Here, we show that the central complement components already existed in the primitive protostome lineage. A functional homolog of vertebrate complement 3, CrC3, has been isolated from a 'living fossil', the horseshoe crab (*Carcinoscorpius rotundicauda*). CrC3 resembles human C3 and shows closest homology to C3 sequences of lower deuterostomes. CrC3 and plasma lectins bind a wide range of microbes, forming the frontline innate immune defense system.

Mammals have dedicated receptors for fungal PAMPs and some of the major surface active receptors are shown below.

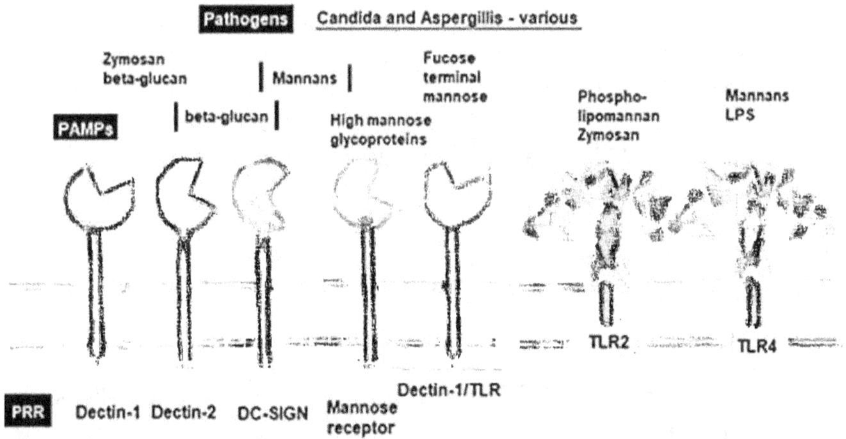

Cell surface receptors that detect fungal residues include Dectin, Mannose, and TLR receptors. Derived from Ramirez-Ortiz and Means.[36]

In addition to TLRs, there are other innate receptors including C-type lectin receptors (CLRs), NOD-like receptors (nucleotide binding and oligomerization domain), and RIG-like helicase receptors (RLRs). A very brief look at each of these receptors gives the feeling that all the pieces of the innate immune system have yet to be fully clarified.

CLRs are widely distributed in the animal kingdom as are LRRs. They are calcium dependent and carbohydrate binding proteins that contain a single or multiple C-type lectin domains (CTDs) that share common folds as motifs. According to Robinson and Moehle, (a) "TLRs are not able to promote internalization of antigens, whereas this is an important function of CLRs." Additionally, (b) "the class-V CLRs are also known as type-II natural killer (NK) cell receptors and include DCAL-1, DCAL-2, and dectin-I. Finally, the class-VI CLRs include mannose receptor (MR) and lymphocyte antigen 75 (DEC-205)." If one looks back at the *Limulus* Factor C molecule (pg. 28) there is a "lectin-like" binding domain within the endotoxin detection biosensor (factor C).

The NOD-like receptors (NLRs) are a group of receptors that serve as intracellular sensors to "see" microbial and danger signals inside of cells. Plants also possess NLRs which give some idea of their ancient origin. Specific NLRs (NOD-1 and NOD-2) activate proinflammatory responses while others form the inflammasome (a multi-protein complex) where caspase is activated. Like CLRs, NLRs are soluble and intracellular receptors.

The "retinoic acid-inducible gene-I like receptors" or RIG-like helicase receptors (RLRs) serve to help distinguish self-RNA from foreign, viral RNA which it does by recognizing subtle differences in viral versus vertebrate RNA. For details and further references for C-type lectin receptors (CLRs), NOD-like receptors (nucleotide binding and oligomerization domain), and RIG-like helicase receptors (RLRs) one is referred back to Robinson and Moehle's original paper (2014) as well as Roitt et al.[37] (2017, page 25) and Monie (2017, page 70).[38]

10. *What are the effector responses of innate immunity?*

Roitt et al. list the following effector mechanisms:
- Phagocytosis
- Complement activation
- Deployment of cells containing cytotoxic granules (Natural Killer cells and eosinophils)
- "Cytokines heavily influence the generation as well as the specific function of effectors within the adaptive immune system."

All but the last item (cytokines) have been discussed in an overview manner.

10.1 Cytokines from PAMPs

Cytokines are the immediate product of the PAMP and TLR interaction. In a nutshell, PAMPs are bound on the LRRs of TLRs and signal through the trans-membrane section thus orienting cellular adapter molecules to signal to the nucleus via NFκB to produce the specific cytokine or cytokines that are pertinent to the specific ingress type as determined by which TLRs are signaling. NFκB, as a nuclear transcription factor, determines which sequences of DNA to transcribe for translation. Like a car engine, multiple TLRs have been shown to act synergistically to increase or magnify the response. An engine firing on 6 cylinders will produce more power than a single cylinder, even though a single cylinder can sometimes produce a very potent response (e.g. endotoxin via TLR4).

Cytokines are divided into those that enhance immune responses on a cellular level, Type 1 (TNF and IFN, etc.) and those that enhance antibody responses, Type 2 (TGF, IL-4, IL-10, IL-13, etc.).

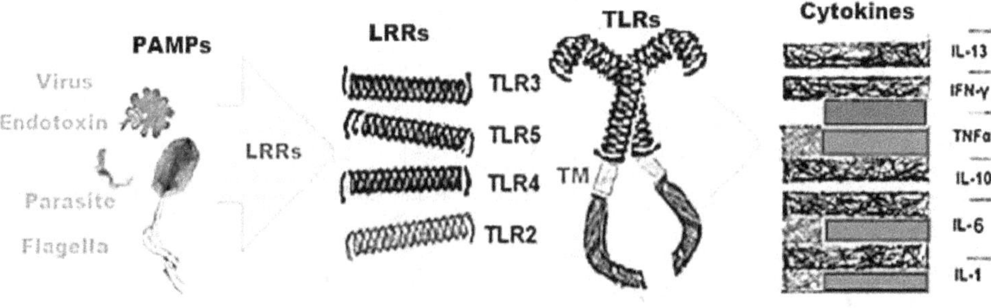

Shown above, a *"cytokine piano"* where the PAMPs on the left are bound by LRRs on TLRs and the resulting conformational change signal is conveyed through the cell membrane (transmembrane domain) and brings about the connection of cytoplasmic adaptors to produce a signal specific for the PAMP that is bound. The "song" that comes out of the keys at right as specific cytokines (interleukins, interferons, tumor necrosis factors etc.) are a product of the notes as given by the specific binding types. Here the notes (cytokines) IL-1 and IL6 and TNF-α are brought about by TLR4 via LPS and INF-γ is brought about by TLR5 via flagella. This specific "song" is denoted by "keys" and cytokine "notes" played (expressed) at right.

10.2 Coagulation

Coagulation preceded any other immune type reaction except perhaps phagocytosis (which has prokaryotic roots), even before there were humoral or separate cellular immune functions as seen in *Limulus* who only has a single blood cell type, neither red nor white cells but only blue amebocytes. In mammals, coagulation provides both an inflammatory response to wounds as well as an internal blood response that can lead to clogging of the microvasculature.

Endotoxin or lipopolysaccharide has a wide-ranging effect in mammals and other metazoans. Fredrick Bang discovered the coagulation of the *Limuli* blood system when he dissected a lethargic juvenile at Woods Hole in mid 1950s. Subsequently, Bang and Levin, a hematologist, were able to work out the mechanism of the reaction and Levin created a lysate of the horseshoe crab's blood that became the long-lasting drug release test reagent (LAL) used to preclude endotoxin from drug solutions and constituents.

The four pillars, as ancient reminants, of systemic response to endotoxin and other PAMPs include cytokine production, coagulation, antibody production and tissue specific reactions. The first two pillars have historically included a wide variety of systemic reactions that includes fever, DIC, sepsis and SIRS. The third pillar includes vaccination, adjuvant effect and ADAs. The fourth pillar includes synergistic effects with other PAMPs, mitogenic-polyclonal activation of lymphocytes, and adjuvant-activation of surface markerts. DIC is disseminated intravascular coagulation. SIR is systemic inflammatory response. ADA is anti-drug antibody.

Given the many effects of endotoxin, in the clinic the needle can swing from fever to coagulation and even disseminated coagulation depending on the source (living and growing etc.) and condition of the host. Of necessity, people have historically focused on a singular aspect of endotoxin's action in mammals based upon a specific test used at the time beginning with the rabbit pyrogen test (RPT, fever) and as superseded by the *Limulus* amebocyte lysate test (LAL, coagulation).

11. Is there a "worst case" PAMP? And how did the PRR against this PAMP develop, ancient to modern, as used in innate immunity?

Viral infections have ravaged humanity from antiquity (see previous vaccination list), thus it would be hard to argue against viral PAMPs as worst-case infectious agents with bacterial infections a close second. Yet, from a PAMP vantage (remember PAMPs are non-infectious), then in terms of potency endotoxins and superantigens (specific bacterial toxins) are worst case. A quick look at each is in order.

What makes endotoxin unique as a PAMP, particularly as a potential contaminant in drug and biologic manufacturing which also carries over to some extent to the clinical environment? The following list forms a simple acronym for endotoxin PAMP characteristics: "LPS".

(a) **Likelihood of occurrence**, LPS forms a dominant structure of the outer membrane of Gram negative bacteria (GNB). GNB are a type that are ubiquitous in water, and existing in all places: fresh, salt water, boiling and even frozen water. Bruce Beutler, Nobel Prize winner, said LPS "has no vertebrate structural analog."[39] If endotoxin is found, it is from Gram negative bacterial origin.

(b) **Potency** is defined by the severity of the response via biological systems. This elicited response is an exaggerated element that complicates clinical treatments and contrasts with the lesser response of every other PAMP except for perhaps superantigens (exotoxins) that are not ubiquitous or heat stable.

(c) **Stability**. Endotoxin requires excessive dry heat applied at >250 C for 20-30 minutes to be destroyed and is more difficult to remove via filtration from aqueous environments that require control such as a large volume parenteral (LVP), small molecule drug (SMD) synthesis or biologics growth-based processes.

Whereas exotoxins are secreted from various microorganisms to poison competing bacteria, endotoxins are an integral component of the GNB cell that are constantly sloughed off in outer membrane vesicles or liberated upon cell death. In mammalian blood systems LPS is then chaperoned to TLR4 via lipopolysaccharide binding protein (LBP) as shown below.

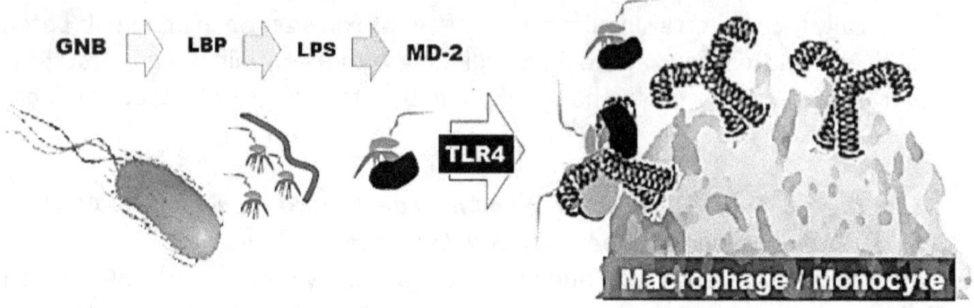

Below, a close up cartoon of TLR4 with prominent MD-2 accessory molecules holding (gauging) LPS in terms of specific fatty acid acyl groups numbers and lengths and saturation as well as core and O-antigen properties. The sixth acyl group (not shown) reaches out from each TLR4 and TLR4*(doesn't completely fit inside MD-2) and is said to be necessary for the most robust response as associated with hexa-acyl lipid A.

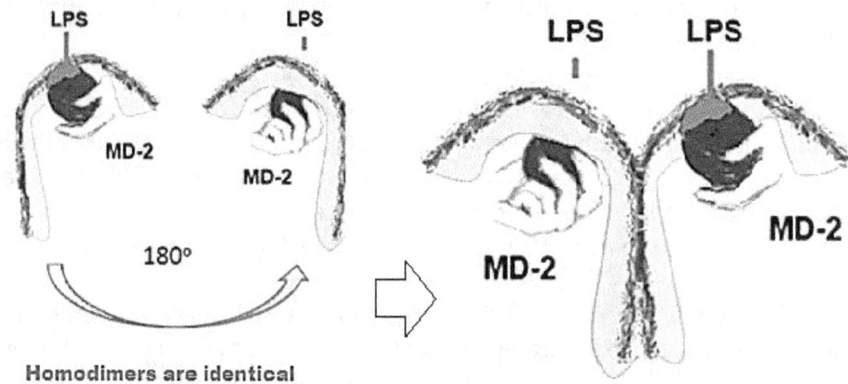

A "handed view" of MD-2 intended to show just that TLR4/MD-2 and TLR*MD-2* are identical but as facing opposite directions (back to back) of the multicomponent complex (LPS/MD-2/TLR4). MD-2 allows for the gauging of each different molecule type in terms of conformational change in TLR5 (below) and volume of filling MD-2 and in binding regions in TLR4 (above).

The cartoon below shows TLR5 and associated bacterial flagellin which is another GNB PAMP. Here binding occurs without accessory molecules such as MD-2 as directly attached to the LRRs of the TLR.

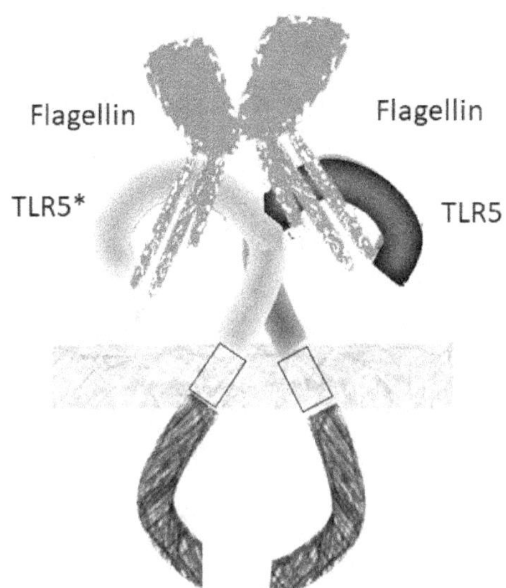

For TLR4, the whole gamut of reactions goes from LPS that is highly potent (agonistic) and reactive to that which is barely reactive to that which actually blocks and negates the response (antagonistic). In mammals, the LPS fine structure produces various responses in different animals. For example, the mouse and horse are responsive to tetra-acyl Lipid IV (a precursor to Lipid A) whereas man is not.[40] This shows the idiosyncratic nature of chaotic evolution and while such innate PRR receptors are highly conserved, they also change over great amounts of time via random mutation and variants exist in the human population referred to as polymorphisms.

Given this "gauging" of response via LPS filling of MD-2 (1 below), an attempt to correlate the most vigorous cytokine response to the worst offenders (I.e. hexa-acyl LPS) results in TLR4 dimerization (2 below), signal transduction across the cell membrane (3 below) varies according to pathway predominately activated (4 below). One leads to a most vigorous response and the other has been associated with an adjuvant-like response seen in vaccinology as well as various "shades of grey" type responses that are mixed signals.

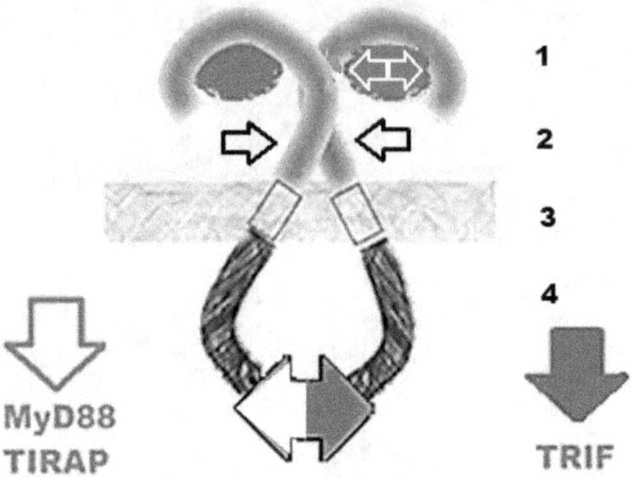

Superantigens are bacterial and viral toxins, often from *Staphylococcus* and *Streptococcus*, and as secreted proteins act as vigorous activators of mammalian T cells. In activating T cells superantigens don't act as a typical antigen that is contained by MHC and presented to the T cell. Rather the superantigen acts to bridge the MHC and T cell and activate multiple T cell types and not specific T cells. Shown below are a normal MHC – T cell interaction and a superantigen interaction in bridging the two structures. Note that more detail around T cells and MHC are coming up in the next chapter.

It is important to note that "superantigens do not require processing to small peptides but act as complete or partially processed proteins. They can bind to major histocompatibility complex class II molecules and stimulate T cells expressing particular T cell receptor V beta chains."[41]

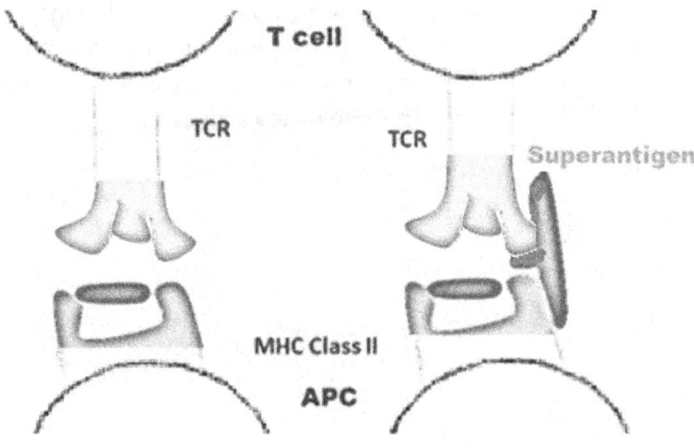

Superantigen outbreaks have included toxic shock syndrome events where the toxins build up in tampons as well as various food-borne illness events. Occasionally a festering wound infection can bring toxic shock.

How has such a complex system developed around endotoxin (as an important example PAMP) detection in mammals? One cannot help but wonder how a complex mammalian system of endotoxin detection has arisen from a much simpler early version. Given the necessity of the activity, no metazoan can long survive without it. Therefore, it likely developed as an overlapping system before completely superseding the existing system. This dynamic feature of evolution where the organism's complexity must change together in an interrelated manner is envisioned below by combining two previous graphics. If we imagine the primitive serine protease system as preexisting and the mammalian system as layered onto it (just as the adaptive system was layered onto the innate system and in much the same way that *Drosophila* Toll has been found to be co-opted for expanded use in TLRs,)[42] then the early stages are more analogous as factor C and LBP actually share some homology- at least in the "endotoxin binding motif".[43] The necessity of cytokine production preceding coagulation gives another layer of protection from inadvertent activation given a much more complex blood milieu (a dozen blood cells, red and white, with hundreds of surface markers and thousands of lymphocyte receptors versus a single cell in *Limulus*).

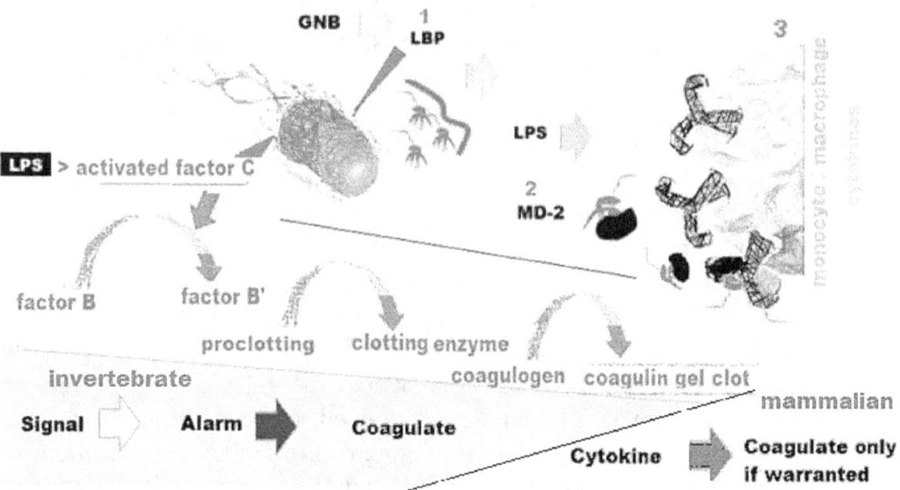

Top, the mammalian response to endotoxin via TLR4. Middle, Limulus clotting cascade. Bottom, some overlap in signaling and a layer added between signaling and clotting in mammals.

12. *What are some of the NEW PAMP types being elaborated?*
The viral and bacterial surfaces are more complex than is often realized. **Surface proteins** on each year's influenza (such as Covid-19) are recognized as the reason that a "one size fits all" approach cannot allow for the development of a single vaccine that endures and continues to work, year after year, for each new strain.

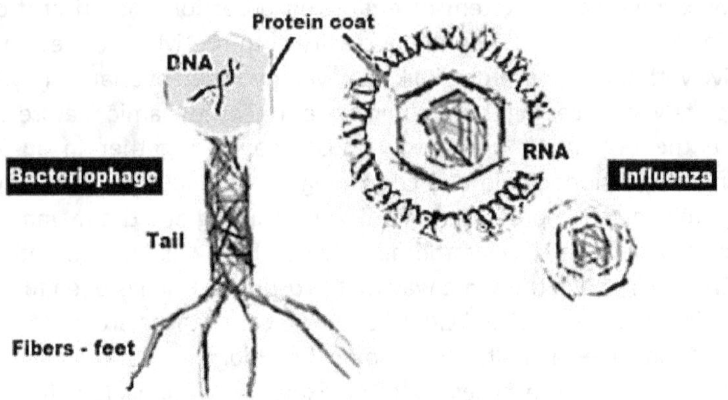

Janeway in *Approaching the Asymptote* talks about the ever-changing nature of PAMPs in influenza surface proteins and the two different systems metazoans have developed over time: innate (ancient) and adaptive (modern) that are specific to generalized receptor structures such as TLRs and also very adaptive as producing "anticipatory" receptors in lymphocytes (BCRs and TCRs).

Infectious agents are highly variable in structure, and their short generation time allows them to alter their structure with great rapidity. This is especially true of their protein structure, which can diversify remarkably even during the course of an infection within an individual. It is not surprising that Macfarlane Burnet, the originator of the clonal selection theory, was studying antigenic variation in influenza viral proteins at the time he fashioned his hypothesis. He realized the necessity for the immune system to develop mechanisms to generate an essentially open and unlimited repertoire of receptors, since evolutionary selection could not provide the immune system with receptors that would recognize the hemagglutinin molecules of next year's strain of virus. Indeed, we now appreciate that the somatic rearrangement of gene segments, including the addition of nucleotides at gene segment junctions, coupled with somatic mutation in antibody variable region genes, provides the immune system with a general mechanism for the generation of diversity that allows the recognition of virtually any molecule by an antibody. Thus, the problem of the generation of diversity has been solved in a unique way in the immune system. The set of germ-line genes has probably been selected over evolutionary time to provide "useful" receptors directed at common pathogens or structural motifs and to provide a broad set of structural possibilities on which to build variants.

Prions are thankfully a very rarely occurring disease-causing agent, however, given the cruel effects they are greatly to be feared as potential contaminants in food stuff from rendered meat (cheap meat that has been mixed with nervous system parts of the animals). The name prion was proposed in 1982 by Stanley Prusiner [44], who later received the Nobel Prize (1997) for the discovery of prions. Prion is derived from the words **pr**otein and infec**tion** and refers to the fact that prions do not traditionally infect but rather "lean upon" other proteins of similar makeup to change their conformation from normal to aberrant. This is a "domino effect" in that once begun it slowly destroys the nervous system in several diseases including "mad cow" disease or BSE (Bovine spongiform encephalopathy) in cows and scrapie in sheep. In humans, a rare form of BSE-like disease has been recognized for years prior to the "mad cow" scare and is called Creutzfeldt-Jakob disease (CJD). BSE is thought to be a rapid acting variant of CJD and is referred to as vCJD. Another type of existing prion disease is Kuru associated with certain cannibalistic tribes of Papua New Guinea.

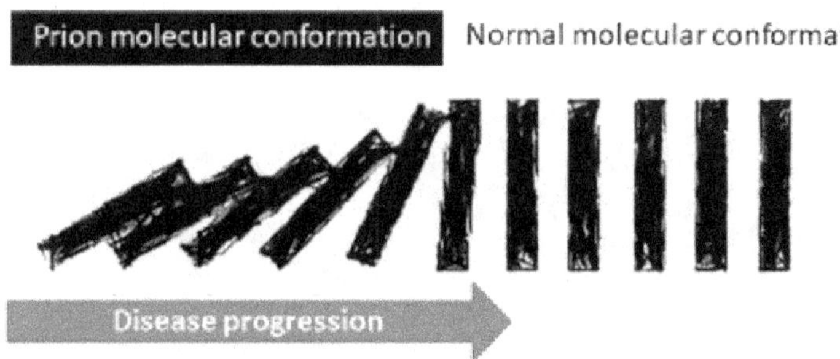

The idea of the prion as an "infectious" particle is shown where disease progression is a function of increasing protein aggregation similar to falling dominos.

It's also very concerning that the diseases of wasting have become quite common in some wild deer and elk (Cervid) populations.

> Unfortunately, the known universe of prion diseases is expanding. At least four novel prion diseases—including human diseases variant Creutzfeldt-Jakob disease (vCJD) and sporadic fatal insomnia (sFI), bovine amyloidotic spongiform encephalopathy (BASE), and Nor98 of sheep—have been identified in the last ten years, and chronic wasting disease (CWD) of North American deer (*Odocoileus specis*) and Rocky Mountain elk (*Cervus elaphus nelsoni*) is undergoing a dramatic spread across North America. While amplification (BSE) and dissemination (CWD, commercial sourcing of cervids

from the wild and movement of farmed elk) can be attributed to human activity, the origins of emergent prion diseases cannot always be laid at the door of humankind. Instead, the continued appearance of new outbreaks in the form of "sporadic" disease may be an inevitable outcome in a situation where the replicating pathogen is host-encoded.[45]

Chapter 3. Workhorse structures & function of adaptive immunity

13. How has the detection architecture developed, ancient to modern, as used in adaptive immunity?

The workings of the adaptive immune system have recently become a highly complex proposition. Here the general knowledge is painted with a thick brush. Only vertebrates have an advanced adaptive immune system beginning with gnathostomes (jawed cartilaginous fish that have a different adaptive immune system also based on extensive rearrangement but by using LRRs). The chart below shows the very basic outline of vertebrates as separated in modern cartilaginous fish and tetrapods.

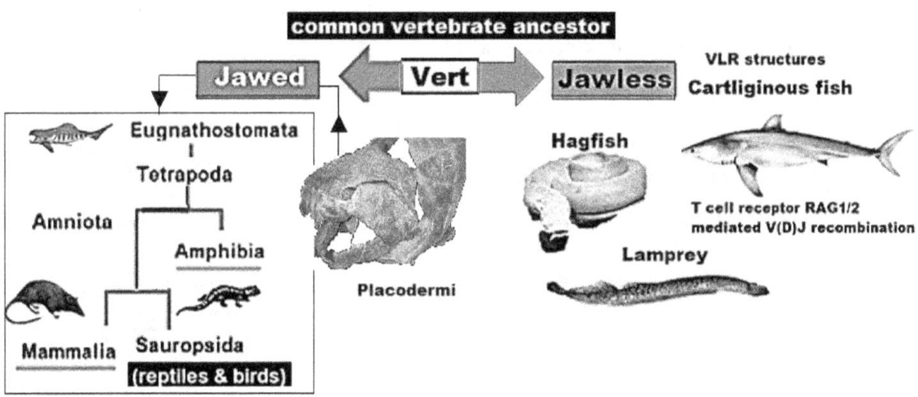

There is no perfect agreement on the evolution of the adaptive immune system. The Placodermi (head shown above) has been identified as amongst the earliest jawed fish (from some 380 million years ago) and became extinct around 300 mya. Modern day lampreys have been studied in terms of adaptive immunity and found to use an adaptive immune system that shares some common features with modern vertebrates while using different structures (architecture). Rather than bony jaws the lamprey have soft toothy suckers. By studying extant lamprey and hagfish, scientists have found that their adaptive immune systems share some variable recombination features with modern vertebrates.

According to Rast and Buckley:[46]

> The adaptive immune systems of vertebrates provide remarkable examples of evolutionary innovation. This is most evident in the unusual mechanisms that jawed vertebrates have invented to create and deploy T-cell receptor and Ig diversity. In a series of papers published over the past decade, the variable lymphocyte receptors of the jawless vertebrates (VLRs) have emerged as an equally powerful example of evolutionary novelty. This story of parallel solutions to the challenge of nonself recognition is made all of the more compelling when one considers the wider system in which these receptors function. Although the V(D)J system of mammals and the VLR system of the agnathans are structurally unrelated, their diversity is expressed and selected in the context of lymphocytic cells that at some level share homology. This similarity offers enormous opportunity for understanding how novel immune mechanisms evolve and are incorporated into a background of more ancient cell regulatory networks.

The lamprey and hagfish adaptive immune system version (VLR) that employs the mixing of gene products (A,B,C) to produce variable receptors and gives another example of using a primitive (LRR) protein motif in a more sophisticated structure to solve immune conundrums whereas vertebrates use immunoglobulin based receptors (TCRs and BCRs and antibody) plus TLRs to a similar effect.

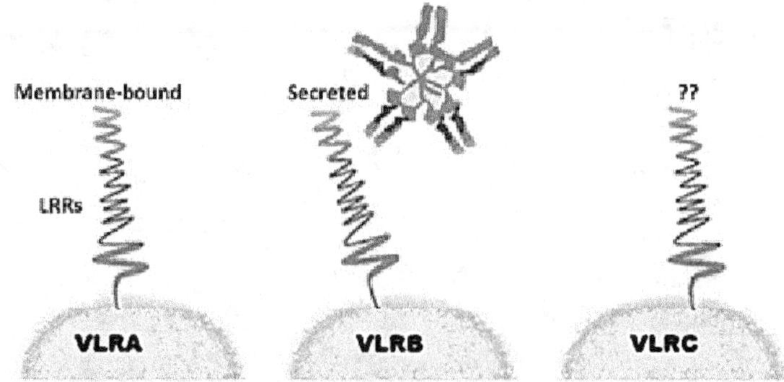

Lamprey and hagfish variable LRR receptors are combined in a "shuffled" manner as are vertebrate V(D)J gene products. Derived from Rast and Buckley.

14. What is the B cell structure and routine function?

A (very) short history of the discovery of B cell in the bird "Bursa of Fabricus" and T cell origins in the Thymus of mammals is given below in an overview graphic derived from Germain.[47] For more detail retrieve the references from the Cooper[48,49] and Miller[50] papers (1965-1968).

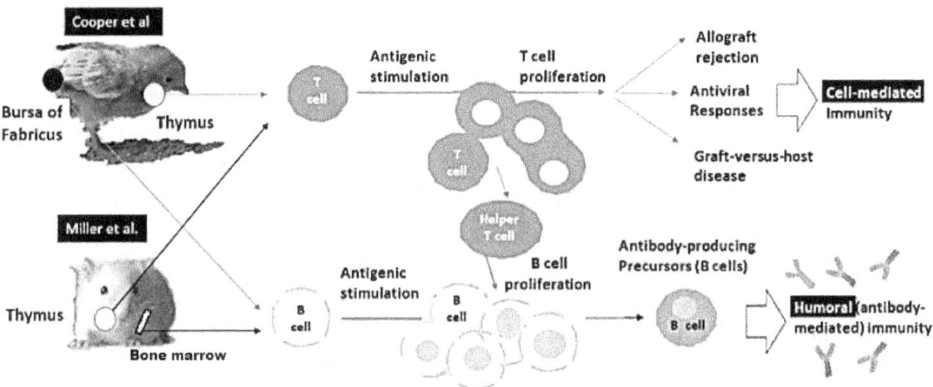

The findings of Cooper and Miller revealed the duality of the major adaptive immune-cell populations as T (Thymus-derived) and B (bursa-derived or bone marrow-derived in birds and mammals respectively) lymphocytes with separate and conjoint responsibilities for host defense.

The B cell receptor (BCR) structure is the same as a secreted antibody structure except in that the BCR has a membrane bound Fc section adapted for transmembrane signaling. According to Templeton and Moehle[51], an antibody is a dimer of dimers, two heavy and two light chains. In a very general way the B-cell receptor appears as shown below with constant regions acting as a universal frame to support the variable binding complementary determining regions (CDRs) also referred to as a "paratope" which is the cognate binding sequence to an antigen's "epitope". The place of binding of the Fab to the antigen is in the cleft between the heavy variable and light variable chains. As seen below there is flexibility in the binding region of the Fab that can be depicted as a range of possible specific binding residues (and is unique to each B cell) that extend from the end to the center of the antibody cleft. This flexibility can be better seen in the variety of detailed graphics given in the Templeton and Moehle reference.

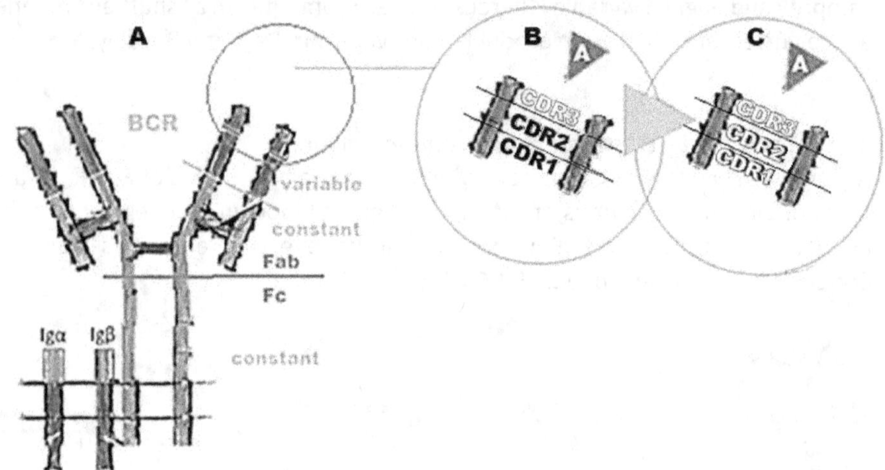

A. Fab and Fc are original chromatographic separation fragments (antibody and crystallizable respectively). Disulfide bonds hold the constant structure together. After all, if the form was too variable then it wouldn't be an "antibody". BCR with variable and constant regions of the heavy (middle long) and light (outside short) chains. **B.** CDRs, at right, shown as 3 regions where, according to Jack and Du Pasquier, "initial RAG recombination only mutates the sequence of CDR3 but leaves the sequences of CRD2 and CDR1 untouched, the repertoire that is produced is far from being as complete as it might be" (pg. 97). These low affinity receptors would not be able to adequately combat a viral or bacterial infection thus, as seen in **C.**, they are directed to lymphoid tissue germinal regions where CDR1 and CDR2 are also "fit" to the specific antigen. "The value of this can be seen by the fact that cold-blooded vertebrates, which lack germinal centres, have problems 'maturing' their immune responses" (pg. 98).

There are several different methods by which antigen is processed by B cells that can result in the production of antibodies. The three methods are described by the three different forms of antigen responded to:

(i)   **Follicular B cells** "express highly specific monoreactive B cell receptors (BCRs), are present in the lymphoid follicles of the spleen and lymph nodes, and typically require T-cells in order to generate high- affinity antibodies, and to undergo class-switching." (Roitt).

(ii) **Thymus-independent antigen binding"** B cells. Antigens such as bacterial cell components like LPS that have highly repeating units can activate B cells polyclonally, which means they don't need to specifically match a given paratope (via a specific epitope). See the figure below, part (A).

(iii) **Thymus-dependent antigen binding B cells.** Primed B cells can present antigen to T-helper cells via MHC II. Antigen bound to the BCR is internalized in endosomes that then fuse with vesicles containing MHC class II molecules. The resulting presentation to T cells will be described in some detail (that includes B cells as an APC). An example of this is that B cells are the dominant APCs that initiate CD4+ T cell response to VLPs (virus-like particles).[52]

We think of B cells as only reacting to external threats via antibody, but here we see that they are able to internalize B cell attachments of a particulate type to contribute through MHC II to cell-mediated versus humoral immunity. There are also "innate-like" B cells (B1 and marginal zone (MZ)) B cells that have polyreactive BCRs of a broader specificity that can recognize some microbial conserved antigens and in this regard are functionally like Toll-like receptors (Roitt pg. 207).

In the prototypical response, a BCR meets a PAMP match (cognate) then it is cloned and spits out many copies of the membrane bound receptor in soluble form. This receptor has been truncated via exon removal (mRNA modification) to produce the antibodies minus the transmembrane sections. The production of antibody proteins has been described by Jack and Du Pasquier as hyper-productive, some 2000 molecules per second!

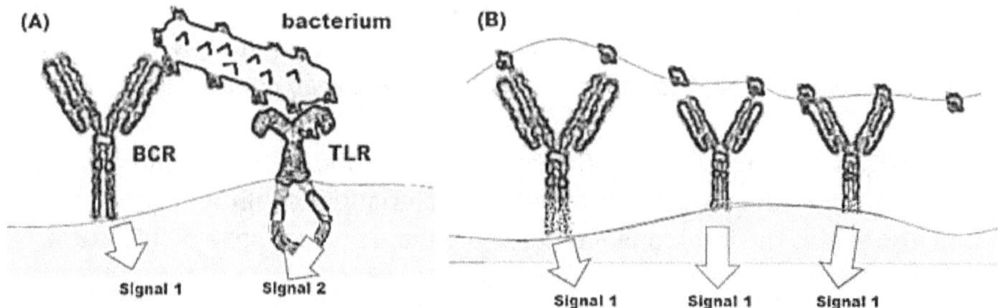

Derived from Roitt et al. figure 7.22. B-cell recognition of (A) type 1 and (B) type 2 thymus-independent antigens. The complex gives a sustained signal to the B-cell because of the long half-life of this type of molecule.

As shown above (B), some B cells do not need T cell (cytokines) to produce antibody, when the BCR are cross-linked by repeat carbohydrate residues found in the cell walls of many bacterial species including LPS and also unmethylated CpG DNA. Other B cell antigens require T cells to become activated.

Antibodies are integrated with both cells of the adaptive immune system as well as with various components of the innate immune system. Note that B cells do contain various TLRs (A above). And complement is active in targeting whole cells and binding to antibody by binding Fc receptors (shown below).

A simplified version of the detection and disposal of a PAMP and the bacterial cell it exists on is shown as a function of antibody and FC receptor binding below. The effector cell includes various phagocytic cells such as macrophage that also possess TLRs.

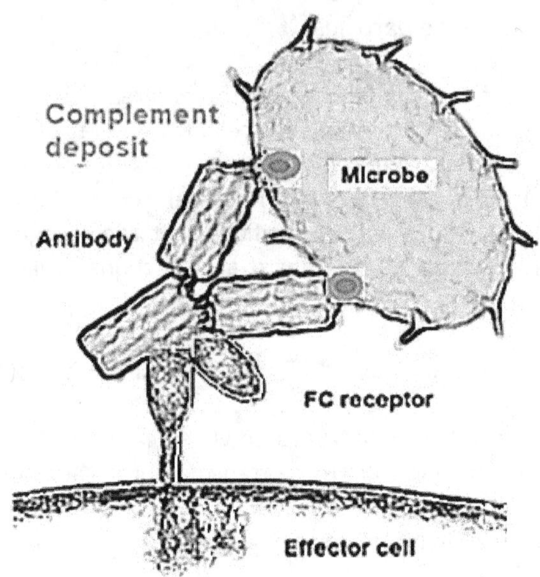

Not to scale as a bacterial cell is much larger than a molecule (antibody). Complement deposited on bacterial cell is bound by antibody which in turn is attached by an FC receptor to an effector cell which can phagocytize it.

The tagging of bacteria with first complement and then by antibody via FC receptor fulfills the purpose of targeting bacteria for engulfment by phagocytes and also shows the interconnectedness of innate and adaptive architecture.

Thymus-dependent antigen binding B cells
The figure below demonstrates the complex activation of B cell via helper T cell in germinal centers where the antigen that is bound by BCR is absorbed into the B cell and is degraded into peptides where each peptide is presented to the T cell via MHC II. This process of forming peptides for APC MHC presentation to T cells will be detailed for T cells in a further section. Cytokines are secreted which in turn promote clonal expansion of the B cell and subsequent secretion of antibody.

Reference the figure below. Compare to Roitt Figure 7.25, the caption from which is used here: Antigen captured by the surface Ig receptor (BCR) is internalized within an endosome, processed, and expressed on the surface of the B-cell with MHC class II. Co-stimulatory signals through the CD40-CD40L (CD154) interaction are required for the activation of the resting B-cell by the T-helper cell. In addition to CD40L-based co-stimulation, helper T-cells also provide additional stimulation to the B-cell in the form of cytokines such as IL-4.

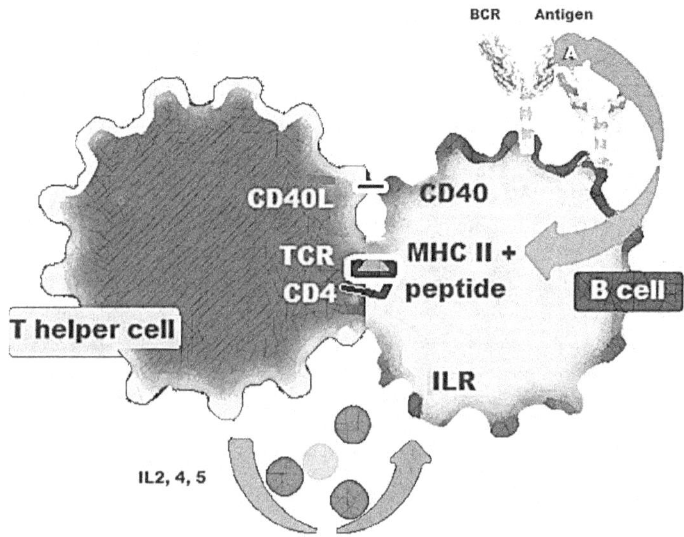

See also graphic on page 65.

15. How is antibody diversity generated?

The difficulty of understanding how the generation of seemingly unlimited antibody binding diversity could be accomplished with a limited set of genes (as the genome is finite) had persisted for decades. References for these quotes are listed immediately below.

> ...Burnet (4) in 1957 termed (the proposed method generating antibodies) clonal selection. This model, now accepted, holds that there is a population of cells in the body that are already differentiated in their ability to bind specific Ag (antigen). Each of these cells bears only one type of Ag receptor and must await a chance encounter with an Ag that happens to fit. The Ag receptors are encoded in the genes of each lymphocyte by a random process that generates billions of receptor specificities before any encounter with Ag. When Ag binds to a receptor, that particular lymphocyte is prompted to proliferate, and each of the progeny bears the same receptor as the parent cell. The question now became how an unlimited number of specificities could possibly be encoded in a limited number of Ag receptor genes.

> This genetic paradox was thrown into sharp relief by the discovery that Ab proteins consist of two regions: one relatively constant and the other highly variable. In other words, Abs gave every sign of being encoded by more than one gene. How could this be, given the one gene, one protein model?

Following evidence rather than dogma, Dreyer and Bennett (5) proposed in 1965 that the V and C regions are the products of more than one gene. Their model was both elegant and subversive. Yet just over a decade later, Susumu Tonegawa's Nobel Prize-winning experiments provided incontrovertible proof. Comparing DNA from mouse embryos with that from myeloma cells (corresponding to differentiated B cells), Tonegawa (6) showed that the genes encoding the V and C regions were separate in germline DNA but close together in mature B cells. The inescapable conclusion was that somatic DNA rearrangement had occurred.[53]

4. Burnet, F. M. 1957. A modification of Jerne's theory of antibody production using the concept of clonal selection. Aust. J. Sci. 20: 67– 69.

5. Dreyer, W. J., and J. C. Bennett. 1965. The molecular basis of antibody formation: a paradox. Proc. Natl. Acad. Sci. USA 54: 864 – 869.

6. Brack, C., M. Hirama, R. Lenhard-Schuller, and S. A. Tonegawa. 1978. A complete immunoglobulin gene is created by somatic recombination. Cell 15: 1–14. 7. Baltimore, D. 1974. Is terminal deoxynucleotidyl transferase a somatic mutagen in lymphocytes? Nature 248: 409 – 411.

Tonegawa was awarded with a Nobel Prize in 1987 for work done in the 1970s and reflected in his 1976 paper: *"Generation of antibody diversity"* (see Roitt page 88).

Yet, one critical issue remained unsolved, namely that the number of possible antibody specificities exceeded the number of genes present in our body. The solution to this was identified by Tonegawa as the rearrangement of gene fragments. This recombination allows the generation of more than one million specificities which further increases numerically by additional mechanisms to up to some 10^9 specificities. Tonegawa was honored with the Nobel Prize in 1987 "for the discovery of the genetic principle for generation of antibody diversity"[54]. Principally, antigen diversity of the T cell receptor is based on similar genetic mechanisms.[55]

This is a highly complex topic, however, the salient points are fairly straightforward theoretically. These are that the genome contains four sets of different genes used to make the antibody molecule and these are called V, D, and J, often designated as V(D)J genes and C which are constant genes. These are germ-line encoded in the different sets, however, they are "mixed and matched" in a much greater variety than the genome could readily store as multiple individual sequences when producing the antibody protein. The mixing and matching is done "somatically" which is to say they are mixed and matched in ways that are not passed down by the genome. This is done using RAG (recombination activating gene) enzymes.

What Tonegawa showed experimentally was that although the V(D)J genes were located on more than one chromosome (separated in the genome on different chromosomes) they are brought together to encode variations for T cell and B cell variable sequences as contained in lymphocyte receptors. Since they were not together on the original genomic sequence, then they had to be rearranged somehow somatically.

The problem in gaining a complete understanding of the source of this hypervariability generating capability persisted in that somehow gene rearrangements could occur somatically by some unknown means that had yet to be elaborated even after the generalized "how" of diversity generation was established, as related by a Brandt and Roth "Pillars of Immunology" paper. References for these quotes are listed immediately below.

It took several more years of work to understand that RAG1/RAG2 are actually the enzymatic machinery of recombination in both B and T lymphocytes (11). Soon thereafter, the Gellert and Mizuuchi (12) laboratories showed that the mechanism of site-specific DNA cleavage by the RAG proteins bore striking similarities to the hydrolysis and transesterification reactions conducted by certain cut-and-paste transposases and retroviral integrases. This astute observation led to the current hypothesis that adaptive immunity evolved through transference of the RAG transposon into an ancestral Ag receptor gene in jawed vertebrates (13, 14).

11. McBlane, J. F., D. C. van Gent, D. A. Ramsden, C. Romeo, C. A. Cuomo, M. Gellert, and M. A. Oettinger. 1995. Cleavage at a V(D)J recombination signal requires only RAG1 and RAG2 proteins and occurs in two steps. Cell 83: 387–395.

12. van Gent, D. C., K. Mizuuchi, and M. Gellert. 1996. Similarities between initiation of V(D)J recombination and retroviral integration. Science 271: 1592–1594.

13. Agrawal, A., Q. M. Eastman, and D. G. Schatz. 1998. Transposition mediated by RAG1 and RAG2 and its implications for the evolution of the immune system. Nature 394: 744 –751.

14. Hiom, K., M. Melek, and M. Gellert. 1998. DNA transposition by the RAG1 and RAG2 proteins: a possible source of oncogenic translocations. Cell 94: 463– 470.

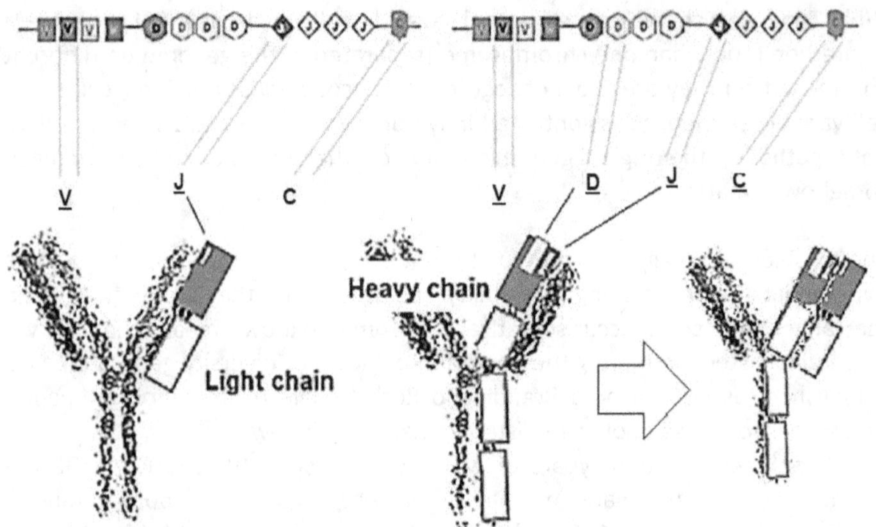

The above graphic is based upon a graphic from Janeway, Travers, and Walport et al.[56] and gives a good generalized picture of generating antibody diversity. The shuffling of gene sets that eventually give rise to very different antibodies shows the formation of the light (left) and heavy (right) chains from V(D)J and C (constant) genes.

Antibodies exist in three forms: isotypes, allotypes, and idiotypes. **Isotypes** are the common form that are encoded by the genome of a given species. **Allotypes** are inherited variants that consist of mostly constant regions of a type per individual inheritance. **Idiotypes** are variants that recognize specific antigen epitopes post somatic rearrangement. This later version is the one that finds and binds antigen and is cloned for mass production in response to verified antigen ingress.

Just as PAMPs are recognized by TLRs and complement, "antigens" recognized by antibody are pieces either on or separate from microbes that are the shapes and symbols recognizable by B cell and T cell receptors and against which antibodies are created. So "PAMPs" are recognized as conserved microbial symbols and "antigen" is recognized as variable microbial symbols. Innate receptors are also conserved to match the conserved symbols and are called PRR while adaptive immune (lymphocyte) receptors (TCR/BCRs) are generally not referred to as PRR, though they are analogous.

Antibodies are only produced by B cells that have differentiated into plasma cells. The BCRs are highly variable due to somatic rearrangement of several (three) sets of genes (called V, J, and D). The structure of each antibody is preserved and defined by "constant" regions and is made hyper-useful via the matching of

60

interfacing variable regions to microbial artifact rearrangements. Thus, constant meets constant in innate and variable meets variable in adaptive (microbe to metazoan).

16. *How does the same membrane-bound BCR become a secreted, soluble antibody? And what is class switching?*

According to Jack and Du Pasquier (pg. 103) in regard to the membrane-bound versus secreted version: "To be secreted the receptor has to be modified so that it is no longer bound to the cell surface... the solution that emerged involves removing the exon containing the transmembrane domain by directed alternative splicing at the RNA level".   The secreted antibody form (left) and membrane bound form (right) of antibody and B cell receptor are shown below as sharing the complementary-determining regions.

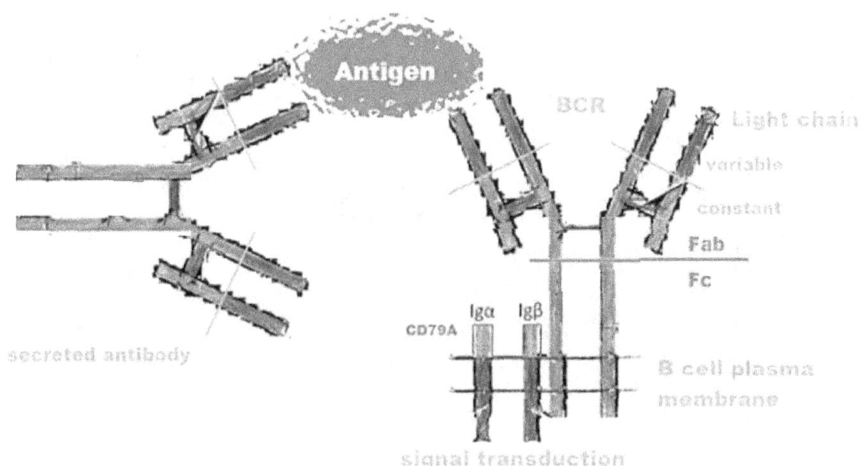

Class-switching is another modification of the basic antibody molecule, in addition to the somatic rearrangement of CDR epitope-binding regions and the formation of a secreted version (antibody). Class-switching is a biological mechanism that changes B cell production from one type to another, such as the isotype IgM to the isotype IgG. During this process, the constant-region of the antibody **heavy chain** is changed, but the variable region of the heavy chain stays the same. Since the antigen binding doesn't change then the antigen specificity is preserved. However, as the antibody has been "lucky" to find its cognate antigen, it can now (via class-switch) interact with different **effector** molecules to facilitate the removal of antigen. The "switching" takes place at the mRNA level in the plasma cell that produces the antibody. Some beneficial "reasons" for class-switching are listed by Jack and DuPasquir.

(a) In some cases it may be best that the receptor activate the complement system,

(b) In other cases it may be better to have the receptor secreted across a mucosal surface

(c) While yet in others it may be important that the receptor can cross the placenta

(d) Each of these responses is best achieved by treating the entire antigen-binding part of the BCR as a modular unit that can be plugged into one of a set of different "constant" regions that define the effector function of the molecule

Class-switch recombination (CSR) occurs in "S regions" which are repetitive and "G-rich on the nontemplate strand, that are found upstream of each CH exon except Cδ." (Roitt et al., pg. 93). In this way the mRNA switches out the constant regions of one antibody to the molecule of another at the mRNA level, as shown below.

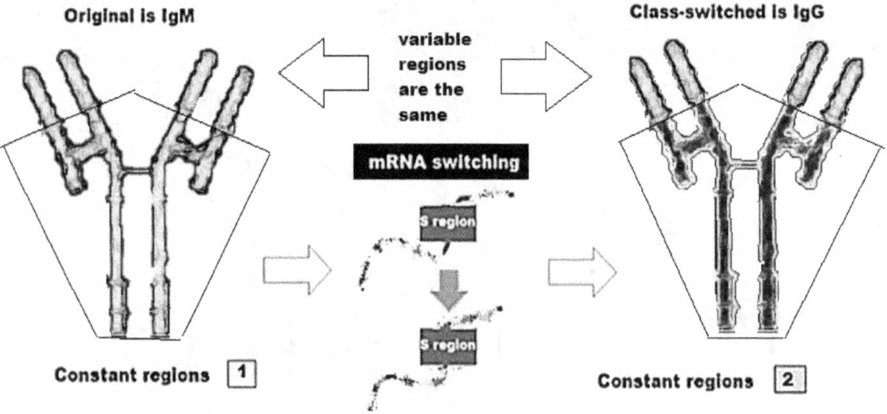

17. How do cells of the organism "know" who is self and who is non-self? That is, how is central tolerance established?

A historically pressing matter has been *"how"* the interfacing of antigens with BCRs and TCRs occurred in the first place. But perhaps the best way to answer this "how" is to answer a related "how" that has to be answered by the body and that is the how of establishing "central tolerance". How can hypervariable receptors (B or T) know the difference between two similar protein peptides? Aren't proteins and peptides just the same old amino acids over and over as used throughout the body? To establish central tolerance, every T cell in the arsenal must be made to be self-tolerant, that is, it must become aware of what the self-proteins look like (or taste like as described below). This is accomplished by parading them through the thymus. Those chewed up proteins (peptides) that don't bind too tightly to MHC I are "accepted" and those that do bind too tightly are destroyed at a significant cost to the host.

The binding of MHC I and T cell is a given as that is the purpose of T cells, however it is the degree of binding that appears to make the difference. To establish central tolerance various proteins in the thymus, all of those in the genome(!), are chopped up via proteases in the proteasome and sent to the endoplasmic reticulum to be joined with a MHC (major histocompatibility complex) that presents such pieces of proteins to T-cells (CD4+ or CD8+).

Jack and DuPasquier[57] (page 95) describe the process of establishing central tolerance. Here this discussion is paraphrased. Thymic epithelial cells are "end differentiated" cells so they only should express a small subset of the entire genome. However, a gene called AIRE licenses these cells specifically in the thymus to express every gene from the genome! This is phenomenal. These then proceed through the proteasome and are presented as peptide MHC-class I complexes to CD8+ cells. This process allows for detection and destruction of those that could be autoreactive. The "parade" to establish central tolerance is imagined below.

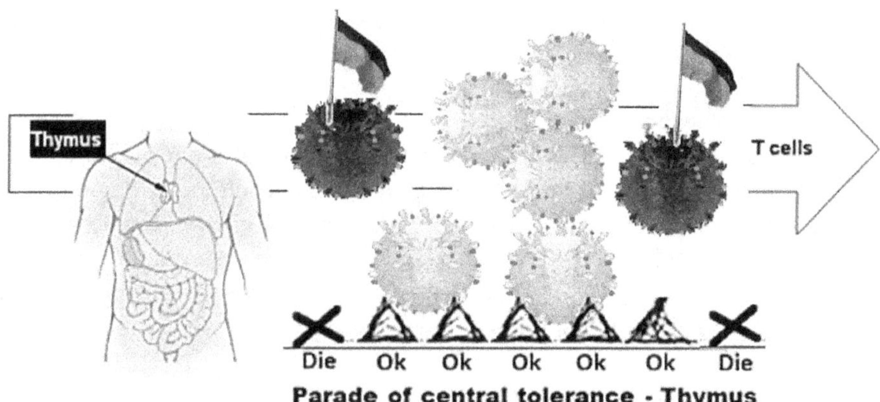

Parade of central tolerance - Thymus

A cartoon *"parade of central tolerance"* by T cells through the thymus. The "ok's" have been deemed to be non-self-reactive whereas the "die's" tagged with a flag have been selected to die because they may be self-reactive. This is not a perfect system and later the confirmation of co-stimulation using non-clonal receptors (rather than these clonal receptors) will be discussed. Motto: *"We can fly 10,000 flags but not 10,001"*.

Interestingly, in the parade of central tolerance it is not as if those T cells chosen to be rejected (to die) do not bind MHC as this is a requirement of all T cells but they rather bind them too tightly relative to those that are selected to conform to central tolerance.

18. What is the TCR structure and routine function?

Central tolerance is established via MHC I and T cells in the thymus, but the general scheme of the process is the same for the detection of peptide antigens via MHC I or MHC II (mainly on APCs) routinely as shown below. Routinely, either MHC I or MHC II are used to detect antigens from internally chewed up proteins (peptides) that are either recognized as self-peptides or non-self as digested from phagocytized cells or particles.

A confusing picture is painted and then clarified by Roitt et al.

> We have just seen how MHC class I presents endogenous antigen while MHC class II presents exogenous antigen. However, between 10-30% of class I molecules present antigen of exogenous origin and a similar proportion of MHC class II molecules present peptides derived from either cytoplasmic or nuclear antigens... The answer to this conundrum lies in the phenomenon of cross-presentation. Phagocytosed or endocytosed antigens can sneak out through channels in the vacuole into which they have been engulfed and thereby gain entry to the cytosol. Once they enter the cytosol they are fair game for ubiquitination and subsequent degradation by the proteasome, followed by TAP-mediated transfer into the ER, and presentation by MHC class I....
>
> Conversely, some of the proteasome-derived peptides within the cytosol, such as those derived from viral capsids, are of sufficient length to make them potential clients for the class II groove and could make the journey to the MIIC. This can occur by a process known as autophagy, in which portions of the cytoplasm, which can contain peptides generated from the proteasome as well as intact proteins, are engulfed internally by structures referred to as autophagosomes. (see Roitt, pg. 154)

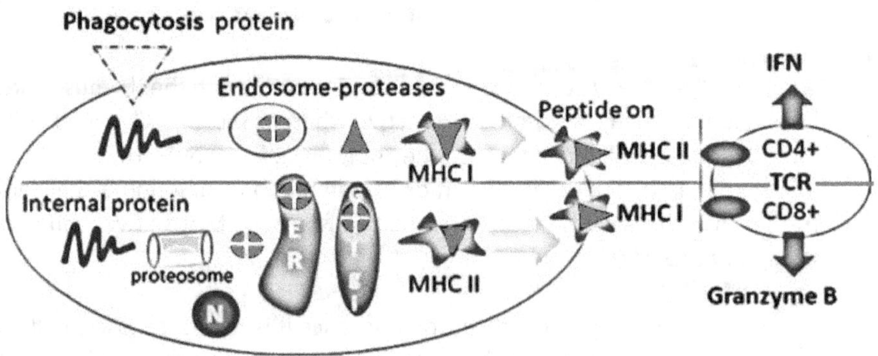

ER is endoplasmic reticulum and N is nucleus. The starting protein (arrow at top) is from phagocytosis of a microbial or viral invader and the protein below it (second squiggle) is from an internal self-protein.   Of course, they are distinguished from self by the TCR due to the previous establishment of "central tolerance". The extrinsic (Class II) and the intrinsic (Class I) proteins are degraded by the endosome and proteasome respectively.   **Granzyme B** is a serine protease most commonly found in the granules of natural killer cells (NK cells) and cytotoxic T cells and IFN is an interferon cytokine often secreted as a response to viral infection.

The overall process indicated above is called "antigen processing and presentation" and is defined as [according to Jiang, Natarajan and Margulies in Jin and Yin (*Structural Immunology*)] "the process of generating antigenic peptides, proper folding of the MHC molecules, loading of peptides onto MHC molecules, transporting to the cell surface, and presenting cell surface-bound peptide antigens to T cells for recognition of foreign and dysregulated antigens...".

19. What is the relationship between HLA, MHC and TCR and the collective function?

The relationship of HLA to MHC to TCR is that of the genes (HLA) that produce the chaperone (MHC) for intracellular proteins that are chopped into peptides that are complexed and presented to T cell receptors (TCRs) as potential foreign antigens (or to verify self-antigens). Human Leukocyte Antigens (HLA) are called "antigens" because they were discovered as a result of organ transplant rejection. What is surprising is the number of HLA genes that provides such diversity of structures for MHC (polymorphic). We saw that one way of providing such diversity avoided using the genome to hold all the genes necessary to provide different structures (antibody CDRs via somatic rearrangement). But the HLA-MHC solution of providing diversity has taken an "expected" route by sourcing directly from a large number of HLA genes as shown below on chromosome 6. The number of HLA genes circulating in the human race has been put at around 500,000 and this diversity is thought to be necessary to aid survival from various historical epidemics.

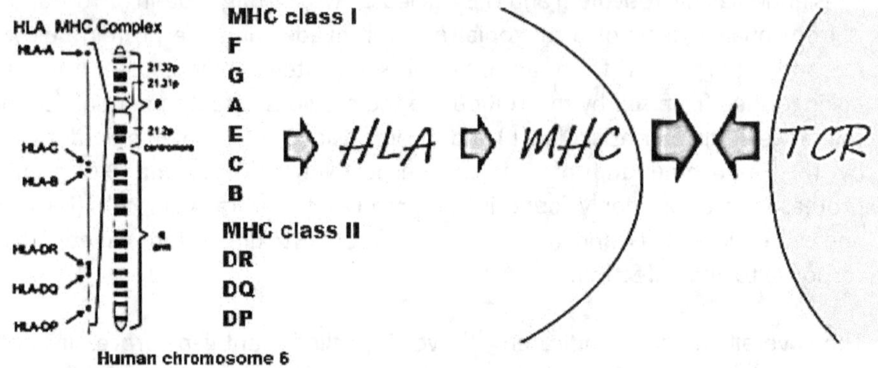

According to Jiang, Natarajan and Margulies (in Jin and Yin, Structural Immunology).

> In the human, major attention has focused on the human leukocyte antigen (HLA) class I molecules HLA-A, -B, -C, -E, -F, and -G, originally defined by amino acid sequence similarities. (HLA-D designates the MHC-II molecules; to be discussed below.) The homologous molecules of the mouse, the major animal model used in immunological studies, are the H2-K, -D, and -L molecules...

> Of more than 1000 pMHC-I structures deposited in the protein data bank (PDB), about one-fifth represents complexes involving HLA-A*02:01, and about 100 of these represent complexes containing unique peptides of length 8–10. Some of these peptides represent fragments from infectious agents such as viruses, others from tumor antigens, and finally some from self-protein...

> Structural comparison of several human MHC-I molecules illustrates that different molecules, by virtue of distinct charge, hydrophobicity, and size of pockets of the binding groove, may select peptides with distinct amino acid preferences at key positions.

While BCRs exist outside the cell for the purpose of binding extra-cellular PAMPs, the main purpose of T cells is to detect infection INSIDE of cells. Yet it too has an outward facing receptor. Jack and Du Pasquier clarify this dynamic.

> ...they must be able to "see" inside each of the approximately 10^{14} cells in an adult human being. How can this be done?... they make use of information extracted from the pattern of proteins that are present within a cell. In a nutshell, this is done by digesting a small fraction of the proteins within a cell into short peptides, which are bound by peptide carrier proteins (MHC), and the resulting peptide-peptide carrier complexes are then expressed on the cell surface, where they may be "seen" by circulating T-cells.

T cell receptors are "Fab-like" Ig structures with three corresponding CDRs per arm like BCRs. Matching MHC molecules, rather than "arms", take on the appearance of cupped hands that hold the MHC-I peptide and present it to the T cell receptor.

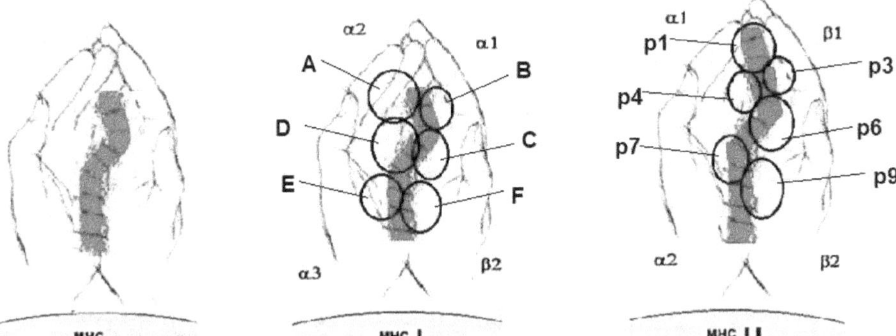

MHC class I holding 8-10 amino acid peptides while MHC II can hold 9-24 amino acid peptides (but most are 13-15) to present to TCR. Cartoon shows the main chains for MHC I and MHC II (α1/α2 and α1/β1 respectively) as well as the 6 CDR binding areas analogous to B cell CDR regions which are termed A-F in MHC I and p1 to p9 in MHC II. Peptide pockets include p1 through p11 with p2 and p10, p11 (for example) making up minor binding pockets and thus not shown. See Jin and Yen pages 27 and 31.

At the risk of adding yet another analogy, the peptide is the *"hot dog"* in the MHC (I or II) "bun". The truncated protein piece (peptide) of 9 (8-10) or 9 to 24 amino acids is the hotdog for MHC I or II respectively. Note that in the figures above and below MHC II has a longer hot dog that extends to the edges of the MHC whereas MHC I is completely confined inside the MHC I pocket. The TCR at right can be said to be "tasting" the hotdog to see if it is self or non-self and if it is non-self, and if it is a cognate structure that it can bind, then it can activate the T cell. If the hotdog analogy is not helpful, the top down look at MHC is shown at right. Like the B cell receptor, the T cell receptor has a cleft (black outline) where CDRs bind the peptide antigen (white). The basic TCR structure remains to be described.

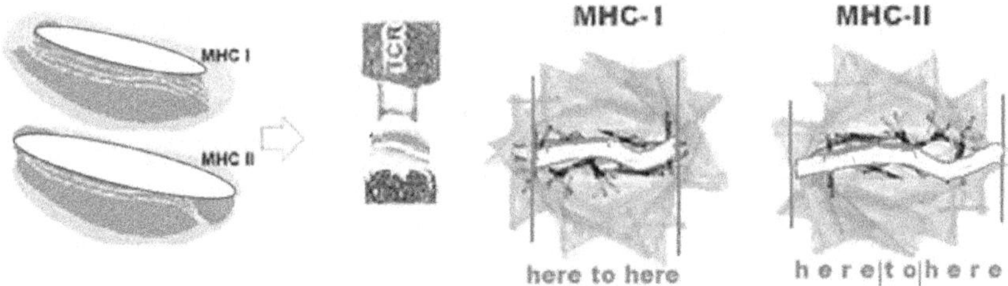

All nucleated cells (most cells except erythrocytes) have MHC-I and are subject to T cell surveillance via the MHC-I-T-cell connection. If MHC provides a peptide that does not conform to the expectation of established central tolerance, then it will be destroyed (such as in the case where such a cell has been hijacked and is making VIRUS proteins). If a self-cell does not have an MHC receptor then it is subject to destruction by cytotoxic Natural Killer cells (NK). As, "one of the main functions of NK cells is to patrol the body looking for cells that have lost expression of the normally ubiquitous classical MHC class I molecules; a situation that is known as "missing-self" recognition. Such abnormal cells are usually either malignant or infected with a micro-organism that interferes with class I expression" (Roitt et al.).

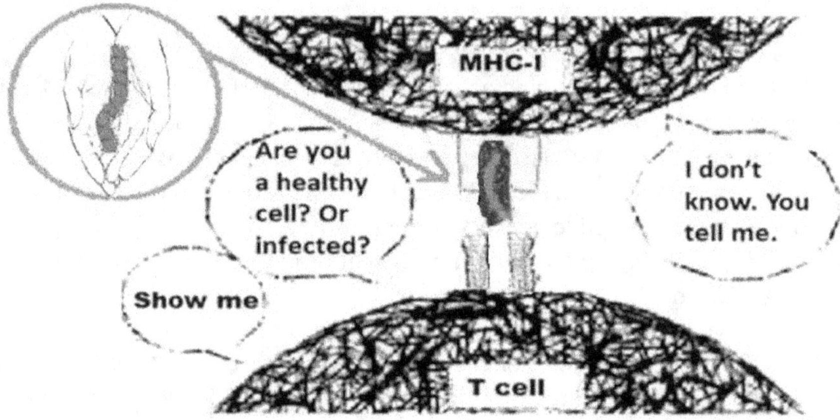

MHC I presenting peptide to T cell receptor for it to gauge "self or non-self".

It may be helpful to view the T cell as initially (in evolution) a "free" receptor that resembled the B cell receptor in function. As per Jack and Du Pasquier (pg. 97), the reason for this is that gnathostomes produce a second population of T cells besides $\alpha\beta$ T cells which are $\gamma\delta$ (gamma delta) T cells. $\gamma\delta$ T cells are "thought to bind intact proteins rather than peptides, and hence do not require MHC for antigen recognition." T cells may have preceded the MHC which is equired today for $\alpha\beta$ T cell functionality (and MHC has no other purpose but to present to T cells). Since B cells are able to effectively do the free binding work in attaching to proteins and microbes, then perhaps this allowed for the change of the T cell function toward internal peptide antigen determination.

Chapter 4. *Adaptive immunity relies on innate as a fail-safe*

20.   *What are surface markers and receptors?*

As "central tolerance" is not a perfect system, some additional means of confirming or denying whether an identified antigen (often a peptide) is "self or non-self" is needed to prevent potential autoimmune reactions (immune system attack on self-cells). Both the BCR and TCR *can't quite be trusted* as they are formed from hypervariable somatic rearrangements. Co-receptor and Co-stimulation / co-inhibition signals are the means of accomplishing the important task of confirming or denying an antigen claim of a foreign substance worthy of an adaptive response (and by whom amongst the many immune cells available). They inform the B cells or T cells or NK cells via various (a myriad of different) cluster of differentiation (CD) receptors /markers in determining "self or non-self" status.

CD markers have been identified by raising antibodies against various suspected surface structures and, once identified, have been labeled numerically post discovery. Examples include many CD markers that are either co-stimulatory or co-inhibitory markers but also many that are of different kinds including receptors (TLRs and CRs) and various ligands including adhesion molecules (necessary to hold immune cells together as they communicate and conspire against microbial visitors).

More than 400 cell surface markers have been identified. CD371 is the highest numbered marker as some markers have multiple variants (making the number greater than 400 but the most recently assigned less than 400). They are "surface markers" because they are antigens for the mAbs used to identify them but they may also be receptors such as TLRs or complement receptors. CD markers have been identified by raising antibodies against various suspected surface structures and, once identified, they are labeled numerically post discovery (e.g. CD28 or CD34). Examples include many CD markers that are either co-stimulatory or co-inhibitory in effect, allowing or disallowing, respectively, forward processing and assigning directionality after an antigen binding event.

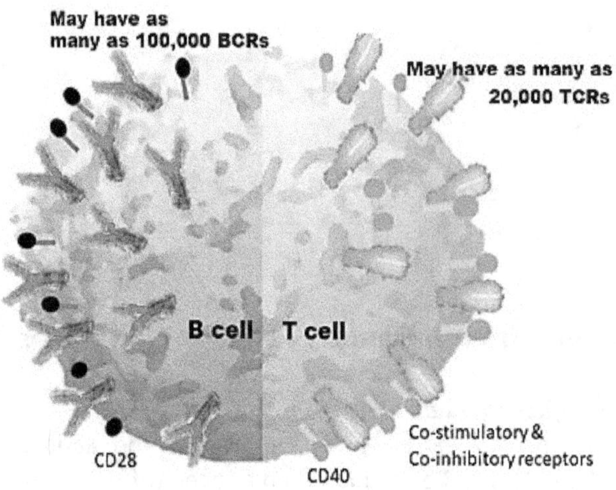

B cell left, T cell right. An example of additional co-receptors include CD28 on B cells and CD40 on T cells.

CD surface markers may be receptors or ligands or adhesion molecules and together the immunophenotyping of immune cells via flow cytometry allow their identification and categorization such as shown below for a B cell. Note that the CD designation such as CD20 identifies both the surface marker and the monoclonal antibody created to bind a specific surface marker.

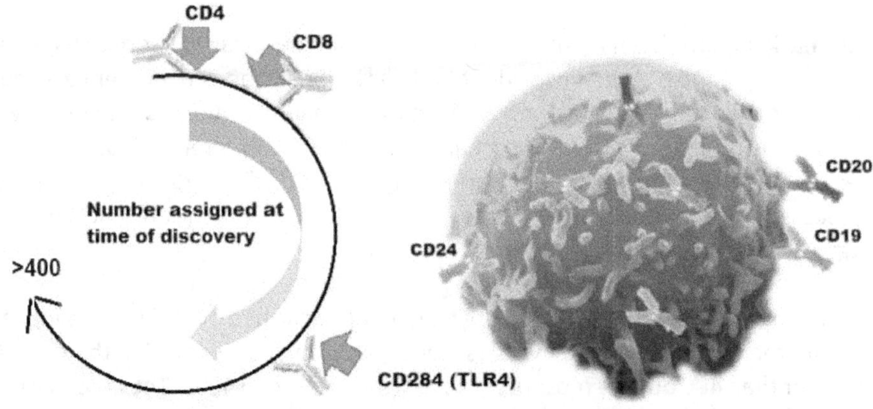

Above left, as surface markers are discovered via sampling immune cell surfaces with various generated monoclonal antibodies and they are assigned a number according to their chronological discovery. The CD number refers to both the surface marker and the mAb used to bind it. The function of the identified CD molecule may not be known until years later. There are just under 400 assigned CD surface markers as of this date. Above right, immune cells are designated by which surface markers they have and which ones they do not have.

In this way, each cell type is assigned an immunophenotype as shown in the table below. CD4$^+$ and CD8$^+$ will be discussed in the next section and are shown in bold and underlined. CD19, CD20 and CD24 are bolded as they are shown in the above B cell graphic.

Cell type	CD surface marker examples
Stem cells	CD34+, CD31-, CD117
Leukocytes	CD45+
Granulocytes	CD45+, CD11b, CD15+, CD24+, CD114+, CD182+
Monocytes	CD4, CD45+, CD14+, CD114+, CD11a, CD11b, CD91+, CD16+
T lymphocyte	CD45+, CD3+
T helper cell	CD45+, CD3+, **CD4+**
T regulatory cell	CD4, CD25, FOXP3 (transcription factor)
Cytotoxic T cell	CD45+, CD3+, **CD8+**
B lymphocyte	CD45+, **CD19+, CD20+, CD24+**, CD38, CD22
Thrombocyte	CD45+, CD61+
Natural killer cell	CD16+, CD56+, CD3-, CD31, CD30, CD38

These CD markers (receptors and ligands) are the ones that monoclonal and small molecules are created to bind as drugs for therapeutic purposes. In cancer it is called "immune checkpoint" when used to prevent immune inhibition by cancer cells. Receptors can be modulated in a stimulatory manner to provoke immune cells against cancer or alternatively inhibitory receptors can be blocked to allow for immune stimulation. Similar molecules (drugs) have been developed for infection and autoimmune reactions.

How are all these markers routinely detected? Via flow cytometry using monoclonal labeled with fluorescent dyes the modern determination of various constituent markers on immune cells can be determined as shown below.

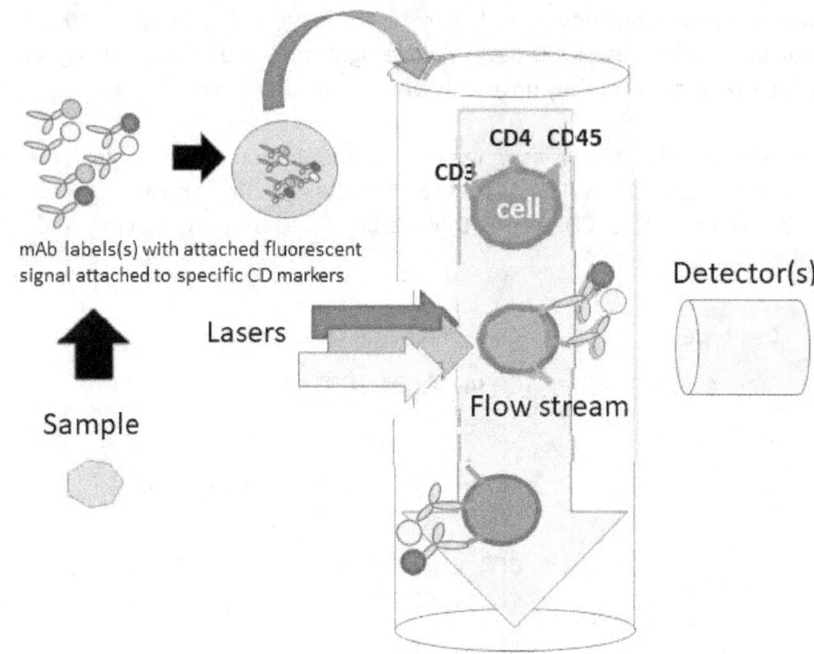

The use of flow cytometry to determine surface markers on immune cells.

A very busy but simple summary of the use of non-clonal receptors (Go/Stop) combined with lymphocyte receptors is shown below.

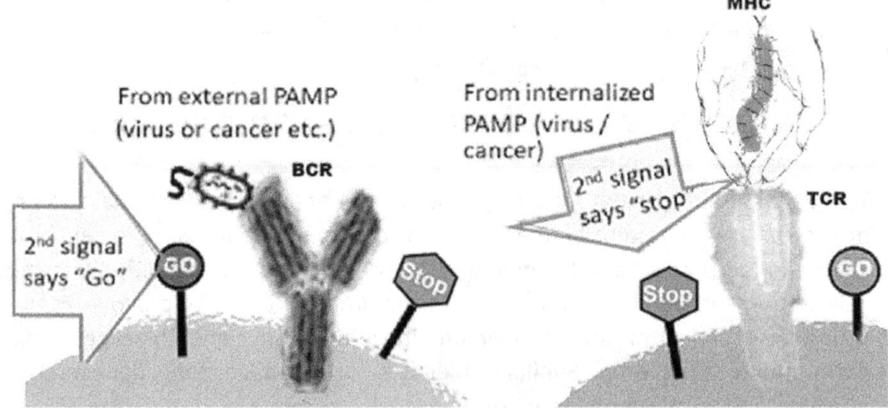

The **B cell at left** can directly interact with PAMPs but also often needs verification via T cell communication (helper T cells). **T cells at right** have two important functions (i) everyday confirmation of "self-versus non-self" for every cell (MHC-I) and (ii) when "non-self" cells are found, to destroy them after detection via killer T cells. MHC-II:T cells detect exogenous rather than internal peptides. The **2nd signal** is confirmation or declination of the **1st signal** (T cell or B cell binding to antigen). At right, MHC presents the chewed up peptide (from a phagocytized cell protein or from an internal cell protein) to the TCR to see if it is self or non-self. It may not be clear that MHC I validation of all nucleated cells is occurring all the time. It is the body's way of distinguishing self from non-self (24/7) regardless of infection.

21. Where does the "second signal" come from?

As Janeway envisioned it, the second signal is needed to "confirm" lymphocyte receptor activation and must come from a perceived microbial source either directly (TLR or CR etc. on a B cell) or indirectly. If indirectly, it comes from CD type markers "induced" by microbial contact. It can also come from the "helper" T cell, a helper to B cells.

> ...antigen-presenting cells (APCs), in particular dendritic cells, are not only uniquely able to capture, transport, and (cross-)present microbial antigens but are also equipped with pattern recognition receptors that provide information about the captured antigen. If this is an infectious agent, CD80 and CD86, the ligands for CD28, the main co-stimulator of primary T-cell responses, are upregulated. If T-cells inspecting the surface of dendritic cells in lymphoid tissues detect major histocompatibility complex molecules loaded with cognate peptide and simultaneously engage CD80/86 with CD28, they are fully activated to proliferate, and under the guidance of additional, cytokine-mediated signals, differentiate into the various types of effector cells; without co-stimulation, they become refractory to further stimulation (a situation called anergy) or even undergo apoptosis. CD28 engagement alone, on the other hand, is without apparent consequence for the T-cells. As we shall see, however, very strong CD28 signals can synergize with the weak tonic signals generated by the process of antigen search itself to trigger T-cell activation. Compared to other, activation-induced, molecules with costimulatory properties such as CD27, OX40, and 4-1BB, CD28's constitutive expression even on resting T-cells explains its central and prominent role in T-cell activation.[58]

And,

> The power of co-stimulated T-cell responses, and the potential damage they can do if misguided, warrants effective control. Cytotoxic T-lymphocyte antigen-4 (CTLA-4) (CD152), which is constitutively expressed on regulatory T (Treg) cells and upregulated on conventional T-cells after activation, is one key element providing this control. This is illustrated by the catastrophic

autoimmune-lymphoproliferative disease experienced by CTLA-4-deficient mice and by autoimmunity observed in cancer patients undergoing CTLA-4 blockade. The intimate interrelation of CD28-mediated co-stimulation and CTLA4-mediated inhibition is apparent not only from their use of the same ligands, for which they compete at the surface of APCs (Figure 1), but also from the absence of autoimmunity in mice lacking both, CD28 and CTLA-4.4 (Beyersdorf et al.).

This story can be exemplified by the T cell CD28 and CTLA-4 co-stimulation/co-inhibition dynamic. CTLA-4 is cytotoxic T-lymphocyte antigen 4. CD28 (co-stimulatory) and CTLA-4 (co-inhibitory) are both modulators of T cell receptor activity and they all act on the SAME three ligands: CD80, CD86 and B7-H2 as shown below.

- **CD28** co-stimulates **T cells** and is expressed on the surface as a 90,000 Da, disulfide-linked homodimer.
- **CD86** is a 70,000 Da glycoprotein (329 amino acids), containing a transmembrane region and a cytoplasmic domain. CD86 is constitutively expressed (always) on APCs (interdigitating DCs, Langerhans cells, peripheral blood DCs, memory B cells and germinal center B cells, and macrophages).
- **CD80** has 10-fold higher affinity for both CD28 and CTLA-4 than for CD86. **CD80** expression is found on activated **B cells**, activated **T cells**, and macrophages.
- **CTLA-4 (cytotoxic T-lymphocyte-associated protein 4**), also known as **CD152**, functions as an immune checkpoint and downregulates immune responses. CTLA4 is constitutively expressed in regulatory T cells but only upregulated in conventional T cells after activation – a phenomenon which is particularly notable in cancers. It acts as an "off" switch when bound to CD80 or CD86 on the surface of APCs (antigen-presenting cells).

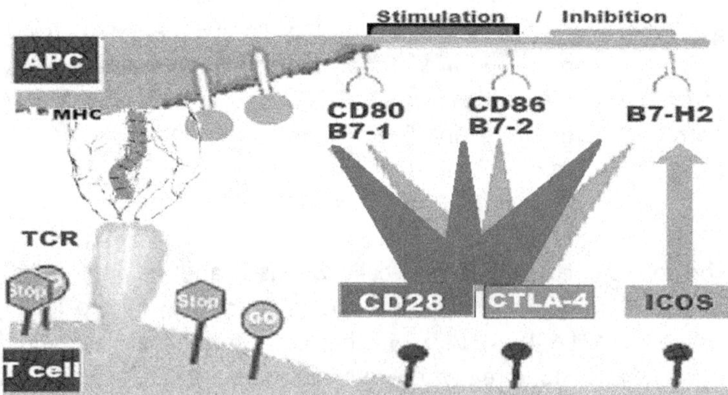

Three scenarios given the competition between CD28 and CTL-4 are given by Beyersdorf et al. below.

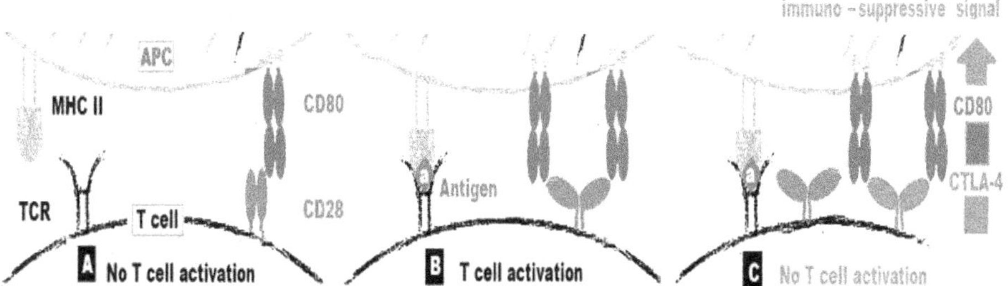

Scheme of the molecular mechanisms underlying CD28-mediated co-stimulation and CTLA-4-mediated inhibition. Notes: (A) In resting T-cells, the lack of an agonistic TCR complex-derived signal together with the suggested propensity of isolated CD28 stimulation to induce inhibitory signaling in the T-cells prevents activation. (B) Upon antigen recognition, CD28 undergoes a conformational change allowing for efficient ligand binding and TCR signal amplification. (C) CTLA-4 counteracts CD28-mediated co-stimulation both intracellularly, that is, in cis, and extracellularly in trans by ligand transendocytosis and activation of immunosuppressive genes in APCs. Abbreviations: CTLA-4, cytotoxic T-lymphocyte antigen-4; TCR, T-cell receptor; APCs, antigen-presenting cells; MHC, major histocompatibility complex. Derived from Beyersdorf et al.

22. What are CD4$^+$ and CD8$^+$ T cells?

CD4$^+$ and CD8$^+$ are T cell co-receptors. They may be thought of as a necessary "secret handshake" between the MHC and the TCR. The co-receptor engagement is necessary as another important validation of the event. It also serves to differentiate the source of the PAMP as from inside or outside the self-cell. The "secret handshake" is shown (crudely) below imposed upon the previous similar graphic. Roitt et al. explain it as follows.

> CD4 and CD8 molecules play important roles in antigen recognition by T-cells as these molecules dictate whether a T-cell can recognize antigen presented by MHC molecules that obtain their peptide antigens primarily from intracellular (**MHC class I**), or extracellular (**MHC class II**), sources. This has major functional implications for the T-cell, as those lymphocytes that become activated upon encounter with antigen presented within MHC class I molecules (CD8$^+$ T-cells) invariably become cytotoxic T-cells, and those that are activated by peptides presented by MHC class II molecules (CD4$^+$ T-cells) become helper T-cells.

This process follows the logic of the lymphocyte's second big question after the first big question (*does this cell contain non-self-proteins?*): **Is this peptide from**

an intrinsic or extrinsic pathogen? And the question is answered by how it is presented, either by MHC I (intracellular) or MHC II (extracellular). In the latter case the cytotoxic T-cell can respond (as an effector) but if it is external to the cell then the T-cell must help the B cell respond (helper T-cell) so it can secrete antibodies against it.

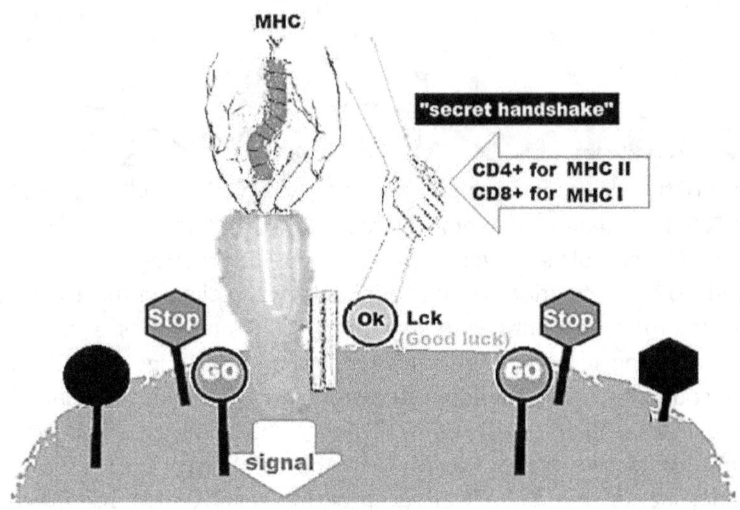

CD4$^+$ T cells are differentiated further according to the cytokine micro-environment, remember that the CD4+ T cells are derived from MHC II where the peptides are presumed from pathogens as sourced from <u>outside</u> the cellular environment.

CD4+T cells are crucial in achieving a regulated effective immune response to pathogens. Naive CD4+T cells are activated after interaction with antigen-MHC complex and differentiate into specific subtypes depending mainly on the cytokine milieu of the microenvironment. Besides the classical T-helper 1 and T-helper 2, other subsets have been identified, including T-helper 17, regulatory T cell, follicular helper T cell, and T-helper 9, each with a characteristic cytokine profile. For a particular phenotype to be differentiated, a set of cytokine signaling pathways coupled with activation of lineage-specific transcription factors and epigenetic modifications at appropriate genes are required. The effector functions of these cells are mediated by the cytokines secreted by the differentiated cells.[59]

As an example of further CD4+ differentiation into Th1 cells to further specify the effector functions as needed based on the interpretation of the threat.

Th1 cells are involved with the elimination of intracellular pathogens and are associated with organ specific autoimmunity. They mainly secrete IFNγ, lymphotoxin α (Lfα), and IL2. IFNγ is essential for the activation of mononuclear phagocytes, including macrophages, microglial cells, thereby resulting in enhanced phagocytic activity. (Luckheeram et al.)

This is borderline overly complex and beyond the scope (of a ~100 page book on all of immunology), yet it fits the generally established mold of immune cell identification of the threat via T cell receptor, CD4$^+$ or CD8$^+$, or B cell receptor and the ability to escalate the response via cytokine, effector activation, and resulting ability to respond to repeat infections (memory response including vaccination to be discussed).

The various accumulating figures shows the compounding complexity (and remains a very partial slice of that complexity) of what is occurring between each antigen presenting cell and T cell encounter. Cytokine expression in its myriad of forms has been referred to as the "third signal".

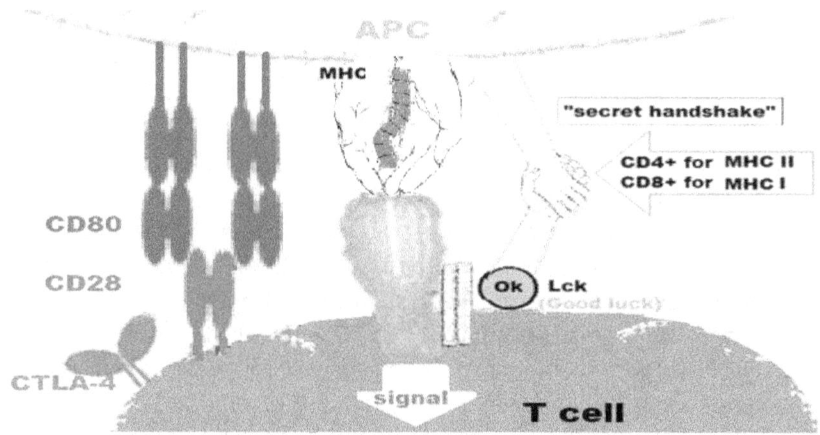

Derived from Roitt et al (Fig. 7.1). Helper and cytotoxic T-cell subsets are restricted by MHC class. CD4 on helper T-cells acts as a co-receptor for MHC class II and helps to stabilize the interaction between the TCR and peptide-MHC (pMHC) complex; CD8 on cytotoxic T-cells performs a similar function by associating with MHC class I.

If you have trouble remembering which CD is associated with what MHC class, then remember that the product of each pair is equal to 8. CD**4** X class II = 8. CD**8** X class I = 8. Also, MHC I harbors intrinsic peptides and MHC II harbors extrinsic peptides. The I can be an "i" and the II could be an "x", with the I's crossed.

The complexity of adaptive immune interactions can be glossed over, as it is here in some cases, to outline the basic functions of each functional part, yet to fill in the myriad of details within the broad outline one must go deeper. The figure below derived from Jack and Du Pasquier gives an indication of this added complexity via the three-way requirement for naïve CD8+ T cell activation into killer T-cells.

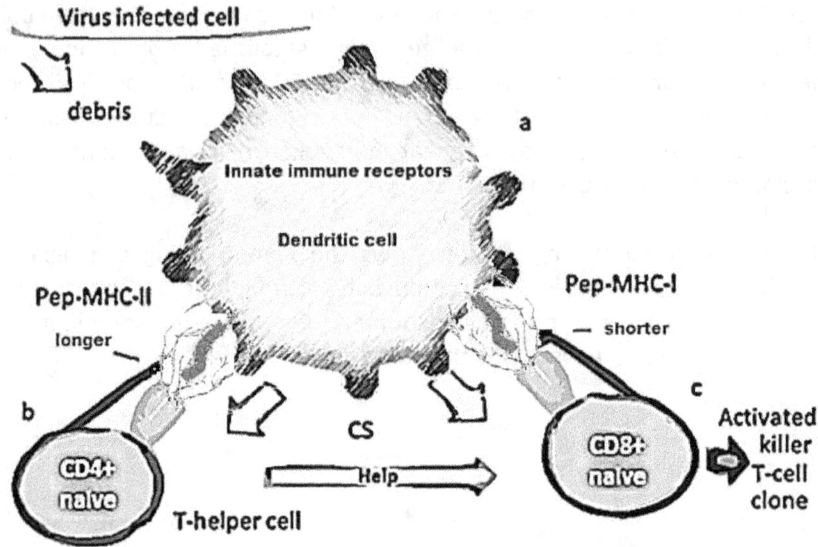

Activation of a naïve CD8+ T-cell. This requires three different cell types-a dendritic cell, a CD4+ "helper" T-cell and the CD8+ T cell. It involves a total of five different immune receptor repertoires: the innate receptors on the dendritic cell; the peptide-MHC-Class-I complexes (pep-MHC-I) displayed on the dendritic cell, the peptide-MHC-Class-II complexes displayed on the dendritic cell, the CD4+ T-cell receptor repertoire and finally the CD8+ T-cell receptor repertoire. A dendritic cell **(a)** detects the danger signals associated with the debris of a lysed virus-infected cell by way of its innate immune receptors (IRs). The debris is phagocytosed and peptides generated from it by digestion in the endosome are displayed on the cell surface as peptide-MHC-Class-II complexes (p-MHC-II) to the T-cell receptor (TCR) of a naïve CD4+ T-cell **(b)**. The dendritic cells also carry on their surface the so-called co-stimulator (CS) molecules, with which they can identify themselves to the CD4+ T-lymphocyte as officially approved "Antigen Presenting Cells". These dendritic cells are also adept at "cross presentation" by which proteins in the ingested material are transferred into the cytosol, digesting in the proteasome and displayed on the cell surface as peptide-MHC-Class-I (pep-MHC-I) complexes to CD8+ T-cells **(c)**. If a naïve CD8+ T-cell interacts with the peptide –MHC-Class I complex, and if it is also assured by the co-stimulatory signals on the surface of the DC that this is indeed a bona fide "Antigen Presenting Cell", and if it also receives appropriate signals as from the CD4+ helper cell, then the naïve CD8+ T-cell will start to divide and produce a large clone of activated "killer cells".

78

They also state that "it is not only CD8+ T-killer cells whose activation is regulated in this way, for a similar sort of control regulates the activation of B-cells… B-lymphocytes… are normally not licensed to autonomously mount a response to their cognate ligand. They require help from T-cells to initiate an immune response."

23. *How is long-term memory established?*

Before going into the current theory of "how" long-term memory is established (albeit somewhat vaguely), an easy "handle" to indicate the presence of a memory cell is surface marker CD27 (CD27[+]) fits in well with previous sections.

> A most useful alternative approach to the identification of antigen-specific cells in the context of known antigen exposures is to use the expression of unique memory-specific surface markers that can be identified by flow cytometry. CD27 was identified as one marker of MBCs[7] and indeed shows many properties that correlate with MBC identity, such as isotype switch and presence of V region mutations. Further, extensive in vitro functional analysis has shown that the CD27+ pool of cells *en masse* behaves differently than the CD27– counterparts. CD27 has also been used to separate human B cells for subsequent gene expression analysis, showing interesting and consistent differences between the CD27+ and CD27– cells.[60]

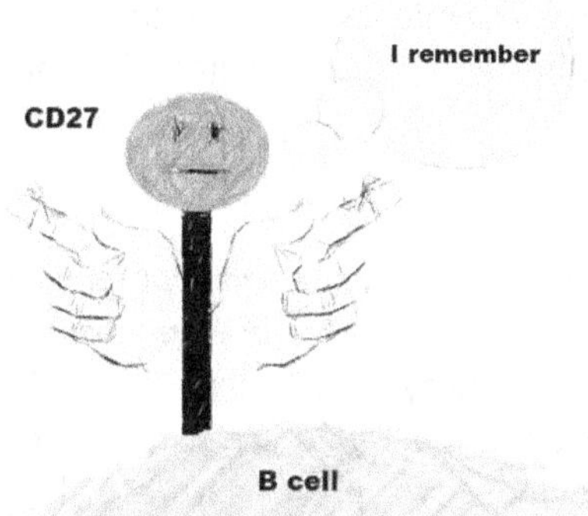

CD27 serves as a marker that identifies likely memory B cells (MBCs)

Roitt et al. gives the general methodology used by the adaptive immune system to "remember" past insults: "…the question arises as to whether the memory cells are long-lived or are subject to repeated antigen stimulation from persisting antigen or subclinical reinfection." And, "Evidence from mouse models strongly suggests that memory T-cells can, at least in principle, persist in the absence of antigen." Finally, "… in humans… on-going entry of new memory cells specific for

---

[7] MBC is memory B cell

diverse antigens to the memory compartment will generate competition between memory cells. ...preexisting receptors of relatively higher affinity in the populations of naïve cells proliferate selectively through preferential binding to the antigen."

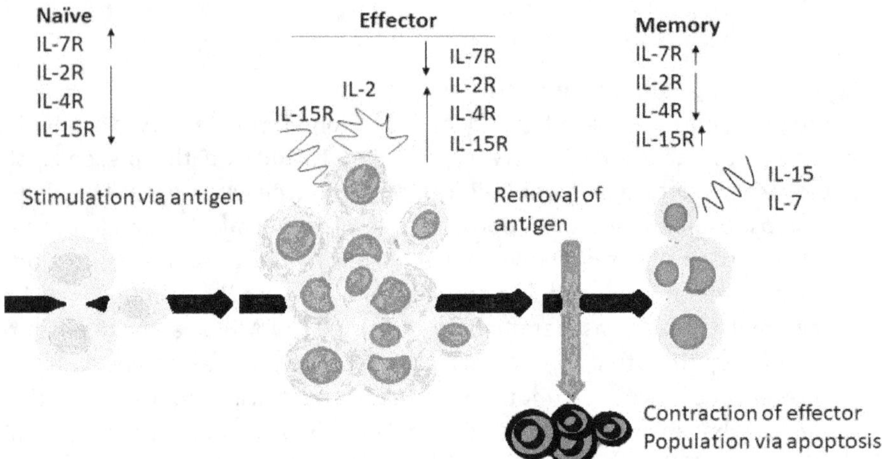

Derived from Roitt et al. (page 267). Arrows show up and down regulation of specific cytokines which influence T-cell proliferation and survival. During resolution of the immune response, massive apoptosis occurs within the effector cell compartment leaving only the "fittest" cells to become memory cells. The memory cell compartment appears to rely upon IL-7 for long-term survival, with IL-15 also thought to be required, particularly for the maintenance of memory CD8 T-cells.

24. What fail-safe practices prevent complex systems from attacking themselves (autoimmunity)? And what happens when this breaks down?
The story as given by Janeway (1989) was that PRRs *must be* responsible for confirming or denying the *"meaning"* of lymphocyte receptor binding events because of the somewhat error-prone nature of highly chaotic somatic V(D)J rearrangement events on top of the imperfect means of establishing central tolerance. Some (such as Tauber [61]) have compared the immunological interpretation of "symbols" and "meaning" to that of the brain, where recognition is somehow tied to receptor activity and the preservation of outside symbols as triggering clues to be recalled upon re-infection. The idea of a "fail-safe" is that if the complex requirements are not met, then the event should fail in a safe manner.

The two lymphocyte receptors (TCR/BCR) are the only mammalian clonal receptors. The CD-type receptors, like TLRs and CRs and checkpoint receptors, are non-clonal in that they are hard-coded in the genome and passed down generation to generation. Mutation occurs of course (as in polymorphisms) but in general this is a high fidelity system that can be counted on to confirm the appropriateness of antigen binding for lymphocyte receptors. Janeway sets a wide backdrop for the understanding of clonal versus non-clonal receptors as a means of interfacing microbial diversity in his 2002 paper (bullets and bolding have been added).[62]

> In this essay, I make four points about the operation of the immune system.
> - First, thanks to the innate immune system's regulation of the main **costimulatory molecules CD80 and CD86**, the immune system rarely mistakes a pathogen for a self-antigen.
> - Second, the adaptive immune system consisting of T lymphocytes and B lymphocytes **can mistake self for non-self because adaptive immunity is selected in single somatic cells**.
> - Third, the adaptive immune system of T lymphocytes and B lymphocytes **is always referential to self**, as it is selected on self-ligands; it persists in the periphery on self-ligands; and at least for T cells, it is dependent on self-ligands to be able to mount a response.
> - Fourth, it is becoming clear that **regulatory or suppressor T cells are our main defense against autoimmunity**... These cells recognize antigen as do all T cells, but they secrete the immunoregulatory cytokines IL-10 and TGFβ.

Though safe guards are included to prevent autoimmune disease, B cell dysfunction still occurs. Autoimmune diseases whose effects correlate with B cell autoimmune-type dysfunction include:
- scleroderma
- multiple sclerosis
- systemic lupus erythematosus
- type 1 diabetes
- post-infectious irritable bowel syndrome (IBS)
- rheumatoid arthritis.[63]

Additionally, many cancers are linked to B cell and precursor malignant transformation including,[64]
- chronic lymphocytic leukemia (CLL)
- acute lymphoblastic leukemia (ALL)
- hairy cell leukemia
- follicular lymphoma
- non-Hodgkin's lymphoma
- Hodgkin's lymphoma
- plasma cell malignancies

T cells have their own disease-associated dysfunction that includes T cell lymphoma and also T cell exhaustion that can occur from persistent antigen insult associated with chronic infection, sepsis or cancer.

25. Why is co-stimulation / co-inhibition a powerful therapeutic paradigm?

Much disease causation is derived from the adaptive immune system going awry, as evolutionarily layered onto the pre-existing innate immune system. Adaptive immunity is very complex and one can envision the many things that can go wrong in terms of disease causation and as involved in all the following disease mechanisms of action (with an associated reference for further study):

(i)   infection[65]
(ii)   autoimmunity[66]
(iii) transplantation outcomes (graft acceptance/rejection)[67]
(iv)   cancer[68]
(v)   genetic interactions and polymorphisms[69]

Cancer, perhaps the least obvious in the list above since it is technically a (transformed) "self" cell, trick T cells into thinking it is not "foreign" by disabling the associated signals that need to be "turned back on" via immune checkpoint therapy. The molecules that perform this therapeutic intervention include monoclonal antibodies and small molecules as well as those changed receptors added to CAR-T cells taken from the patient's own T cells (autologous) or in a generic form developed previously (allogenic).

The co-stimulation / co-inhibition paradigm allows for the man-made manipulation of these tiny levers to modulate the outcome of immune interactions. This knowledge, combined with the advent of monoclonal antibodies has allowed for the pharmacological manipulation of co-stimulation and co-inhibition to battle cancer, infection, and autoimmunity.

The thing about cancer drugs is that there may be three or four or five combo regimens as the tumors suffer treatment but come back as the cancer develops resistance to each therapy.[70] So new checkpoint inhibitors are constantly being developed. The co-receptor paradigm therefore is part and parcel of the most significant new paradigms in cancer treatments including "checkpoint immune inhibitors", the identification of new surface markers, and "CAR-T therapies", all to be discussed from a therapeutic drug perspective in the next chapter.

Chapter 5. The biologics revolution

26. What is a biologic drug?

Biologics are both old and new drugs. Arguably, the earliest biologic was Pasteur's use of a rabies derived concoction as a vaccination-like treatment for rabies which was a shocking accomplishment given the level of scientific understanding of the day. He didn't "mass produce" the vaccine. But he did produce custom lots and provided it for afflicted children including some from the US. The production of vaccines also became a focus of the Pasteur Institute (*Institut Pasteur)* founded in 1887. The institute has produced ten Nobel Prize winners for medicine and physiology since established in 1908. "Germ theory" was still being established and there was no knowledge of viruses. Vaccinology was born and began to be increasingly developed to cover many of the very worst diseases of the early twentieth century (rabies, smallpox, tetanus, polio, etc.). Pasteur had only just disproven "spontaneous generation" which purported that microbes arose *de novo* rather than from other microbes. He used the famous experiment with the "swan-necked flask" to disprove this idea. Also, surprisingly, by the beginning of the 20th century, there was some forming of an understanding of the adaptive immune system and the use of antibody generation in animals (horses, sheep) which employed the inoculation of these animals and subsequent use of their plasma (sera) to treat various infectious diseases such as diphtheria.[8]

---

[8] Analogous to what Pasteur had done in rabbits for rabies and Jenner with cowpox inoculates for small-pox prevention.

What we now think of as a "biologic" is a drug formed from **recombinant technology**. The new "human" form of recombinant human insulin began a new paradigm and was no longer an "animal insulin" which needed to be harvested from tons of cow and pig pancreases but a "human insulin" with the exact proper amino acid sequence. Since then the biotech revolution has been nothing less than breathtaking in scope and curative transformation. Most recently a third and fourth wave of advancement has occurred (the first two being (i) recombinant proteins and (ii) monoclonal antibodies) in (iii) manipulation of immune markers and receptors in immune checkpoint inhibitors and CAR-T therapies to treat cancer and in (iv) gene therapy and anti-sense treatments, all to be outlined here.

What makes a drug a biologic then? It is produced via a living organism, and, as a modern drug is a recombinant product. Molecular complexity and size are also important characteristics. Generally biologics are too complex to be chemically synthesized (as are small molecule drugs) and must be grown up in cell culture reactors. Below, as per Ganellin, Jefferis, and Roberts[71], size and complexity are important and produce the associated technological difficulties in various production methods:

> One has only to consider the size of biomolecular drugs to recognize that the technologies that give rise to biomolecular drugs must be considerably different from the classical SMDs. Genentech equates the difference between aspirin (21 atoms) and an antibody (~25,000 atoms) to the difference in weight between a bicycle (~20 lbs) and a business jet (~30,000 lbs.).[72]

27. What is the difference between a biologic developed in 1900 and 2000?

The difference in the *old and the new* biologics lies in the level of understanding of the underlying scientific principles (none versus a lot) and, importantly, the modern use of recombinant (DNA) technology to "clone" natural proteins and to put them into single celled organisms (or cells from animal tissues such as CHO or plasmids within cells) DNA and produce them in a big cell culture reactor.

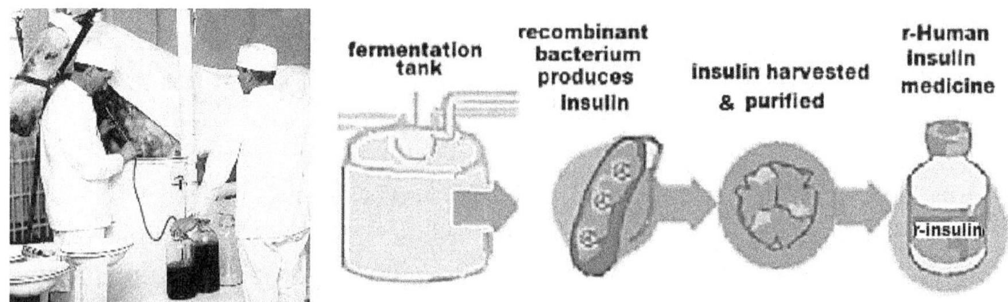

The treatment for diphtheria (left), developed in the late 19th century, relied on injecting horses with a toxin, harvesting their serum which contained (polyclonal) antibodies to the toxin. This contrasts with modern methods employing the insertion of recombinant genes into single-celled organisms who subsequently express the proteins into the cell culture matrix from which it is purified and packaged into a drug (right).

Behring and Kitasato below describe (1890) the budding use of animal serum to treat specific infections.

> Publishing in the *Deutsche Medicinische Wochenschrift*, Emil Behring and Shibasaburo Kitasato described "[t]he mechanism of immunity in animals to diphtheria and tetanus," explaining how infected animals could be cured and healthy animals could be pretreated to prevent infection. Although the article did not explain how animals were immunized, it reported that blood removed from the carotid artery of immunized rabbits and allowed to stand until serum formed could then be transferred into the abdominal cavity of mice to confer protection against subsequent challenge with either tetanus bacilli or tetanus toxin. Furthermore, the serum could be injected therapeutically to rescue infected mice from the lethality of tetanus. Importantly, control blood and serum from non-immune animals did not transfer protection. In a follow-up article, Behring described a similar mechanism by which blood or serum taken from rats rendered immune to diphtheria toxin was sufficient to protect guinea pigs from sickness induced by a subsequent injection of the same toxin. Collectively, the articles demonstrated that the serum of immune animals is capable of neutralizing toxins; that this property is exclusive to immune animals; and that serum can be used to both prevent infection and treat infection.[73]

28. What is the central dogma of molecular biology that lead to the recombinant revolution?

28.1. Recombinant technology

The genome of every organism consists of discrete genes along the linear length of DNA (linear if it were unwound) and these genes are responsible for gene products that include transcription factors, enzymes, structural proteins, and all the other substances of the cell that makeup tissues. The central dogma of molecular biology is simply the posit that genes are *transcribed* via messenger RNA and *translated* by ribosomes into proteins.[74] The "tools" necessary to match and expand upon Watson & Crick's central dogma have been developed over the years and now include a number of important ways to manipulate RNA and DNA and protein. We can look at some of the necessary associated early advances by looking into Arthur "Kornberg's refrigerator".

Kornberg's refrigerator at Stanford as described in Mark Jones' article contained the following cutting edge items early in the "biologics revolution".[75]

1. *Polymerase* to synthesize long chains of DNA and fill in gaps.
2. *Ligase* to join contiguous ends of chains.
3. *Exonuclease III* to remove obstructive phosphate groups at chain ends.
4. *Phage I exonuclease* to chew back one end of a DNA chain.
5. *Terminal transferase*, an enzyme (from the thymus gland) to add nucleotides, willy-nilly, to the other end of a DNA chain.

Humans have learned to utilize and manipulate the transcription/translation process to make proteins for therapeutic purposes that are different from the original source of DNA of the cell to which they are transferred. To insert new sequences into a DNA vector (plasmid or single-celled organism), **restriction endonucleases** (REs) are used. REs are **enzymes** that cleave DNA into fragments at or near specific recognition sites within genes known as **restriction sites**. A wide variety of different REs are available to allow for very specific changes to be made in adding and subtracting DNA from existing structures. REs are derived from bacteria and virions whom have used these enzymes for eons to manipulate host genomes for pathogenic purposes.

A couple of example types are EcoR1 and DNA digestion with **EcoRI** produces "sticky" ends (AATT). Eco stands for *E. coli* from which it was isolated and RI designates that it was the first RE isolated from *E. coli*.

$$\text{G | A A T T C}$$
$$\text{C T T A A | G}$$

Using the **Sma I** restriction enzyme produces "blunt" ended cleavage fragments.

$$\text{C C C | G G G}$$
$$\text{G G G | C C C}$$

One can gain some appreciation for the RE process as used to splice in a given gene by looking at the molecular structure that includes DNA.

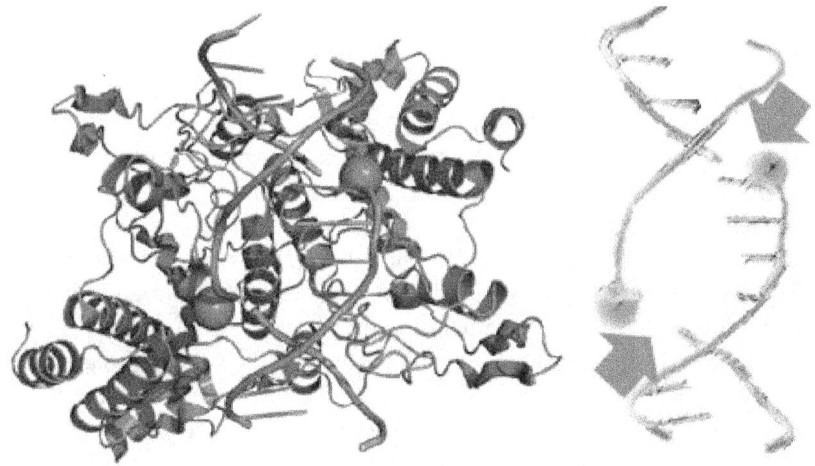

Structure of the homodimeric restriction enzyme *Eco*RI bound to double stranded DNA. Two catalytic magnesium ions (one from each monomer) are shown as spheres and are adjacent to the cleaved sites in the DNA made by the enzyme (depicted as gaps in the DNA backbone). At right, DNA structure contained within the first picture with obscuring features removed. Arrows added to denote cuts in DNA backbone.

The first recombinant drug, recombinant human insulin, was made by inserting the human insulin gene into a single celled organism (*E. coli*) to express the protein into cell culture in giant bioreactors and from there the protein could be isolated and purified into drug quality form for injection.

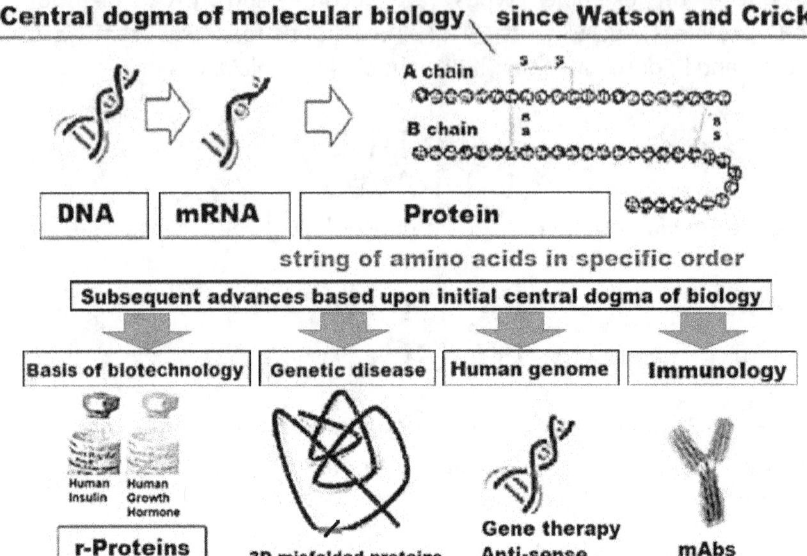

The first commercial recombinant protein was human insulin (Humulin®) manufactured by Eli Lilly & Company in 1982. Human growth hormone soon followed. The discovery of insulin was a godsend for diabetics in the early 1900s and was used successfully as harvested from animals for decades. But the use of animal derived proteins required processing tons of organs (pancreas) from cows (bovine insulin) and pigs (porcine insulin) and pituitary from human cadavers (for human growth hormone). The initial replacement of naturally harvested animal proteins with recombinant proteins established a long-lasting paradigm that not only endures but that has served as a platform for improvements in natural molecules. Rather than just insulin now there are long acting, short acting, fast acting insulins, etc. This powerful paradigm is shown below.

E. coli was historically the first cell used to produce recombinant proteins (given the use of it as an early genetic model organism), which required extensive process purification from endotoxins inherent in Gram negative organisms. The very basics of the historical use of first animal insulin and then human insulin using recombinant protein production in lieu of naturally harvested proteins is overviewed below.

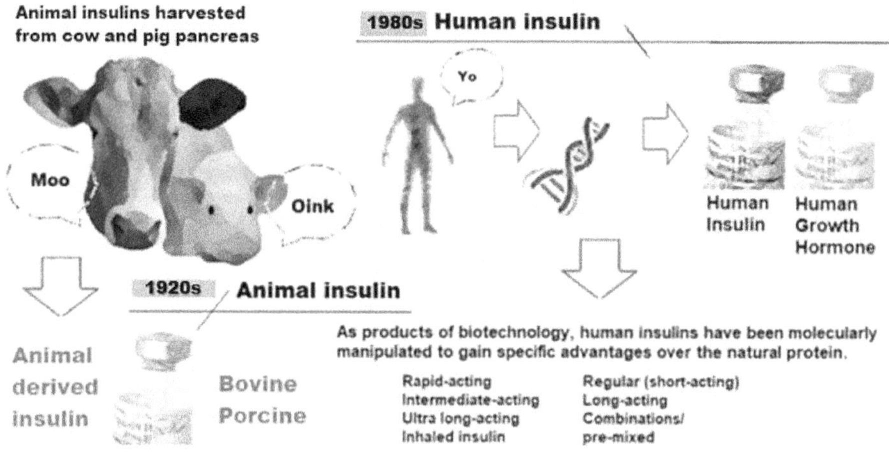

It should be mentioned that, over the decades, the central dogma has been elaborated to include many complexities in the understanding of genes and genomes, such as the understanding of introns and exons in animal genes.

The figures below give an idea of the complexity of eukaryotic and prokaryotic genomes. The genome is all the available genes (DNA sets on the linear, ATCG sequence, but wound in a helix) as presented in chromosomes for eukaryotes (animals) and in nucleated DNA in prokaryotes (bacteria).

An interesting difference in prokaryotic versus eukaryotic genomes, is the presence of **introns and exons** in eukaryotes and the absence (generally) of such a structure for genes in prokaryotes. In the human CRP gene there is an intron separating the signal peptide and the coding region. In *Limulus* this intervening sequence is missing.[76] This is another "advancement" of modern, complex animals over primitive ones (such as arthropods) that has necessitated advancements in knowledge for producing various recombinant molecules. It presents a complication of the original "dogma" of "one gene equals one protein". The dogma remains true but it is also true that one gene may produce more than one protein.

This added complexity presents some hurdles in understanding what is going on. Some of the introns and exons that populate mammalian gene transcription products (mRNA) prior to translation are shown below and give the general idea of the role introns and exons serve.

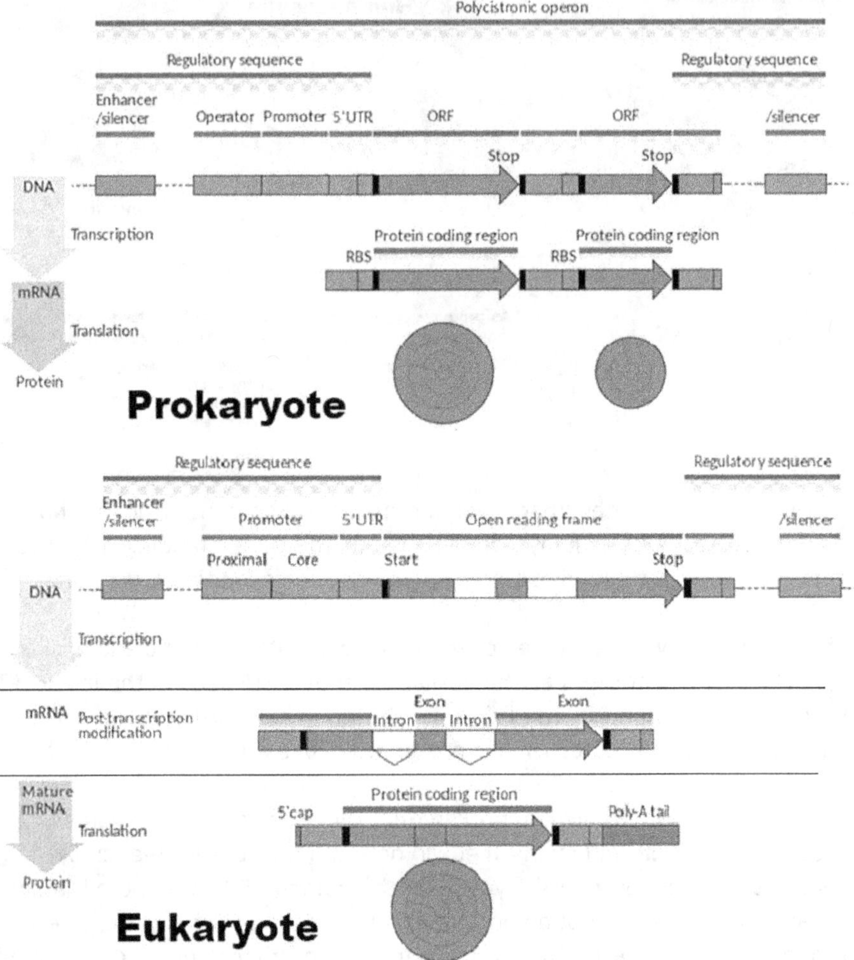

From Thomas Shafee - Shafee T, Lowe R (2017). "Eukaryotic and prokaryotic gene structure". *WikiJournal of Medicine* 4(1). The major difference in complexity is the addition of exons/introns in eukaryotic genomes which is "post-transcription modification". Labels added (prokaryotic or eukaryotic). Greyscale added.

This added complexity built into the mammalian genome presents both additional opportunities for disease (exon /intron mutation) and difficulty in deciphering the linear code. Humans have 46 chromosomes (23 pairs) as shown below. Each chromosome is a continuous strand of double-stranded DNA.

A couple of different views of the intron/exon dynamic are shown below. The first figure from WHO[77] shows how the non-coding regions (introns) are folded out for excision.

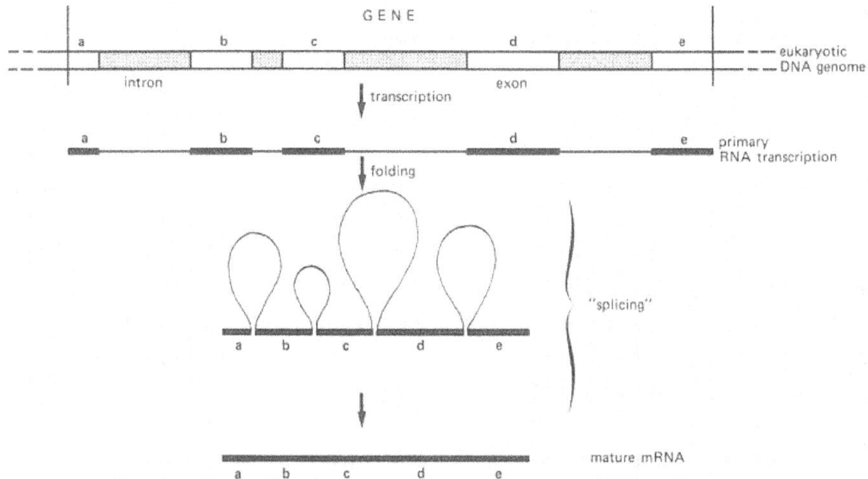

The organization of a eukaryotic gene and processing of the RNA transcript. The DNA sequence coding for a protein, indicated by regions a-e, is interrupted by non-coding sequences, called introns. The regions of the primary RNA transcript corresponding to the introns are excised and adjacent exon regions are spliced to form the mature mRNA, which is translated in the normal way.

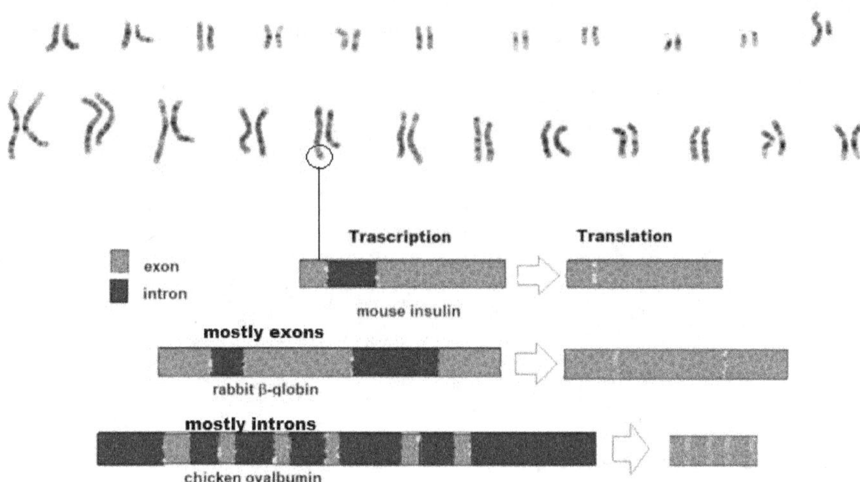

The **human genome project** (HGP) was completed around 1999 but still left about 8% of the genome un-decoded in telomere regions (terminal ends of the "x"). Key findings of the project (2004) are described below.

1. There are around 22,300[78] protein-coding genes in man, similar to other mammals.

2. The genome has many more very similar, repeats of DNA (segmental duplications) than previously suspected.[79]

3. At the time of the draft publication (2001) fewer than 7% of protein families appeared to be vertebrate specific.[80]

It should be obvious that the HGP has been instrumental in promoting the burgeoning sciences of immunology and genetics and evo-devo (study of how animals evolved) etc.   In the previous disease causation list (infection, autoimmunity, genetic defects and polymorphisms, transplantation outcomes, and cancer) one can see that knowing the genetic "intent" (generic human code) versus reality (individual specific mutation or polymorphism) may reveal practical ideas for cure (drugs or corrected genes via gene therapy).

28.2. What are monoclonal antibodies and how are they produced?

After vaccines and recombinant proteins such as insulin and growth hormone, the next major advancement in biotechnology molecule types was the monoclonal antibody. A previous chapter went into detail describing antibodies in general but below is added the historical overview of the technological means of mAb production. Polyclonal antibodies were discussed previously such as those used in the production of diphtheria anti-toxins (polyclonal antibodies). Monoclonal antibodies on the other hand are the product of a single B cell (plasma cell) and the gene that makes it is homogeneously expressed by the expression vector, usually a CHO cell (Chinese hamster ovary), as shown below.

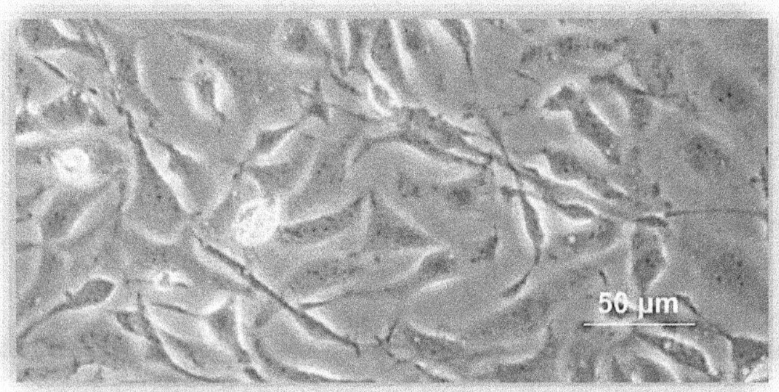

CHO cells are eukaryotic (mammalian even) rather than prokaryotic cells (bacteria) so they are able to produce complex biologics such as monoclonal antibodies which have complex glycosylation (signaling sugars) requirements which as we have seen is only possible in eukaryotes. Not only the complexity but also the size of mAbs are much greater than small early proteins such as insulin (MW ~150,000 vs. ~5,800 Daltons) that were produced in *E. coli*.

Seemingly small improvements can provide significant but unheralded platforms to produce new technologies to manufacture advanced products. Many biotech companies have accumulated proprietary advantages that they rely on to produce their product candidates. Some, like CHO, are universally available rather than proprietary as they were developed relatively early on.

CDRs determine the antibody or mAb binding to various microbial (prokaryotic) cells for infectious protection or binding to surface marker targets on self-cells in the context of cancer and autoimmunity. In the microbial world, as you can imagine, there are millions of potential structures that can serve as antigen.

The wide diversity and hypervariability of antibodies is needed to anticipate these structures as year to year they can change (such as a new flu virus surface protein). This adaptive immune system ability to "adapt" is different from the innate immune system's constant use as a sentinel to detect and dispose of common motifs (PAMPs such as endotoxin or flagella etc.). CDRs come in 3's and the hypervariability is supplied by the somatic shuffling of germ line encoded genes as shown previously.

Monoclonal manufactures have developed sophisticated ways of manipulating the CDR regions to develop novel molecules that can bind specific targets, even targets that they may not naturally bind. Targeting surface markers of cancer cells will be discussed later. The manipulation of CDRs is beyond the scope of this discussion but should give some appreciation of the on-going efforts.

The means of developing monoclonals was an amazing advance from efforts that started before 1975 and culminated in the cultivation of hybrid cells that were immortalized (live forever in cell culture repeat transfers without an animal that they are a part of).

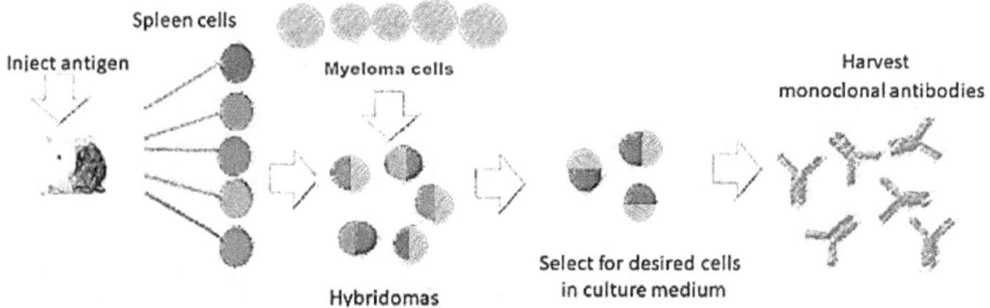

With mAbs Ehrlich's dream in the early 1900's of producing *"magic bullets"* was accomplished. Some of the first mAbs were combined with toxin proteins to deliver a toxic dose to cancer cells. In 1975 Kohler and Milstein made the first fusions of myeloma cancer cells with B cells to create immortal hybridomas that could secrete antibodies against a single (monoclonal) antigen. They received the Nobel Prize for this in 1984. In 1988 Winter's team developed techniques to "humanize" mAbs to reduce or eliminate many side effects that were encountered when using the mouse generated mAbs. Most recently, 2018, Allison and Honjo received a Nobel Prize for using mAbs against surface markers to overcome cancer cell inhibition of the body's immune defenses against cancer.

Roller bottles growing cells that produce monoclonal antibodies for scale up reactors are shown below. Typically, the cells used to grow and express the monoclonal antibody proteins (mAbs) are CHO or Chinese hamster ovary cells.

29. What are some of the newest forms of biologic therapeutics?

29.1 Immune checkpoint inhibitors

Surface marker and receptor science has been described in a previous chapter. Briefly, surface markers are antigens (self-structures not microbial) that can be identified via monoclonal attachment. In this way they have been identified as surface structures and used to identify the cells upon which they reside. Since there are ~400 identified thus far they are used to differentiate cells such as $CD8^+$ and $CD4^+$ T cells which means T cells that use MHC I or II antigen receptors. Differentiation is based upon function, lineage, etc. CD(+) means the cells contain the CD type but CD(-) means that it does not.

Checkpoint receptors are used by immune cells to give a **"go" or no-go" signal to T cells and B cells**. Since the lymphocyte receptors (B and T) are somatically rearranged and thus hypervariable then they need some permission from the hard coded markers/receptors (such as TCRs). An example of this is the most important pair to be discovered thus far as used in cancer treatment. CD28 and

PDL-1 will be elaborated below. Thus, a "checkpoint inhibitor" is a permission giving or denying receptor that gives context for lymphocyte receptors (TCR/BCR).

CD28 (Cluster of Differentiation 28) is a protein expressed on T cells that provides co-stimulatory signals required for T cell activation and survival. T cell stimulation through CD28/TCR provides a potent signal for the production of various interleukins (especially IL-6).

CD28 is the receptor for CD80 (B7.1) and CD86 (B7.2) proteins. When activated by Toll-like receptor ligands, the CD80 expression is upregulated in antigen-presenting cells (APCs). This points to an important distinction between markers that are induced (by PAMPs ultimately) versus those that are constituent (always there). The CD86 expression on APC is constitutive (always there as opposed to activated). CD28 is the only B7 receptor constitutively expressed on naive T cells. Association of the TCR of a naive T cell with MHC:antigen (as discussed in a previous chapter) complex without CD28:B7 interaction results in a T cell that is anergic (non-responsive).

PD-L1 also has an appreciable affinity for the costimulatory molecule **CD80** (B7-1). Engagement of PD-L1 with its receptor **PD-1 (programmed cell death)** on T cells delivers a signal that inhibits TCR-mediated activation of IL-2 production and T cell proliferation. The graphic below is from Gong et al.[81] and shows the very basic mechanism of action for checkpoint inhibitors.

From the Gong paper we can see that Pembrolizumab and Nivolumab are PD-1 inhibitors and Atezolizumab, Durvalmuab, and Avelumab are PD-L1 inhibitors. The first checkpoint inhibitor was developed in 2014 but it is still an active area of research around both new PD-1/PD-L1 and new (different) target markers and receptors as well. In case it's not obvious, a specific mAb can attach to either the receptor (PD-1) or the receptor ligand (PD-L1) to block the signal.

PD-1/PD-L1 inhibitors (of immune activation) have been approved for several different cancer types:

   (i)   NSCLC, non-small cell lung cancer
   (ii)   HNSCC, head and neck squamous cell carcinoma
   (iii)   MSI-H, microsatellite instability-high
   (iv)   RCC, renal cell carcinoma
   (v)   HCC, hepatocellular carcinoma
   (vi)   UC, urothelial carcinoma
   (vii)   MCC, Merkel cell carcinoma

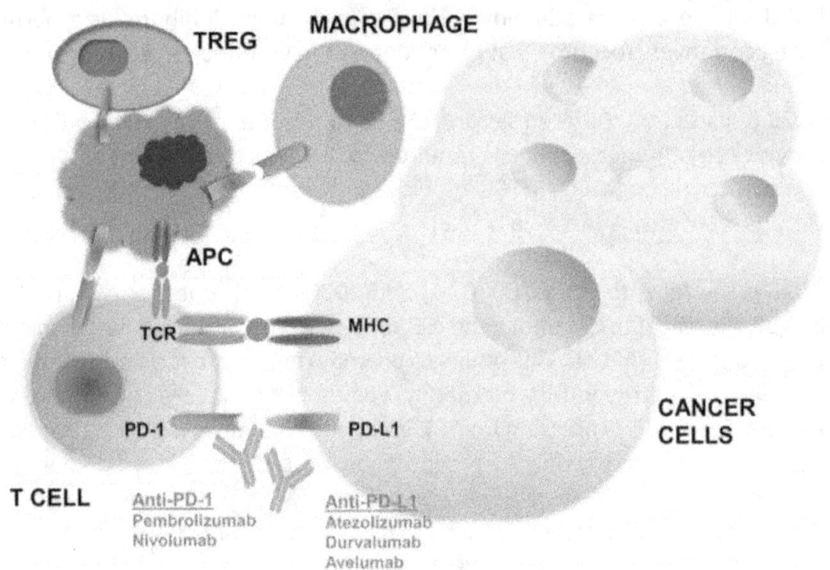

Mechanism of action of PD-1 and PD-L1 inhibitors. The programmed cell death 1 (PD-1) receptor is expressed on activated T cells, B cells, macrophages, regulatory T cells (Tregs), and natural killer (NK) cells. Binding of PD-1 to its B7 family of ligands, programmed death ligand 1 (PDL1 or B7-H1) or PD-L2 (B7-DC) results in suppression of proliferation and immune response of T cells. Activation of PD-1/PD-L1 signaling serves as a principal mechanism by which tumors evade antigen-specific T-cell immunologic responses. Antibody blockade of PD-1 or PD-L1 reverses this process and enhances antitumor immune activity. TCR, T-cell receptor; MHC, major histocompatibility complex; APC, antigen-presenting cell. From Gong et al. (http://creativecommons.org/licenses/by/4.0/). Changed to grey scale.

A couple of different monoclonal antibody types can be used to block the co-inhibitory signal given through PD-1/PD-L1. By blocking this signal with a mAb targeted to either the PD-1 (receptor) or PD-L1 (receptor ligand) the T-cell can respond against the cancer cell. An upcoming chapter describes further the rationale for both T cells and co-stimulatory receptors that make up the "immune checkpoint inhibitors".

Beyond the singular consideration of "immune checkpoint inhibitors" there are also a host of related efforts to engineer T-cells culminating in various CAR-T cell therapies, as summarized in the following figure.

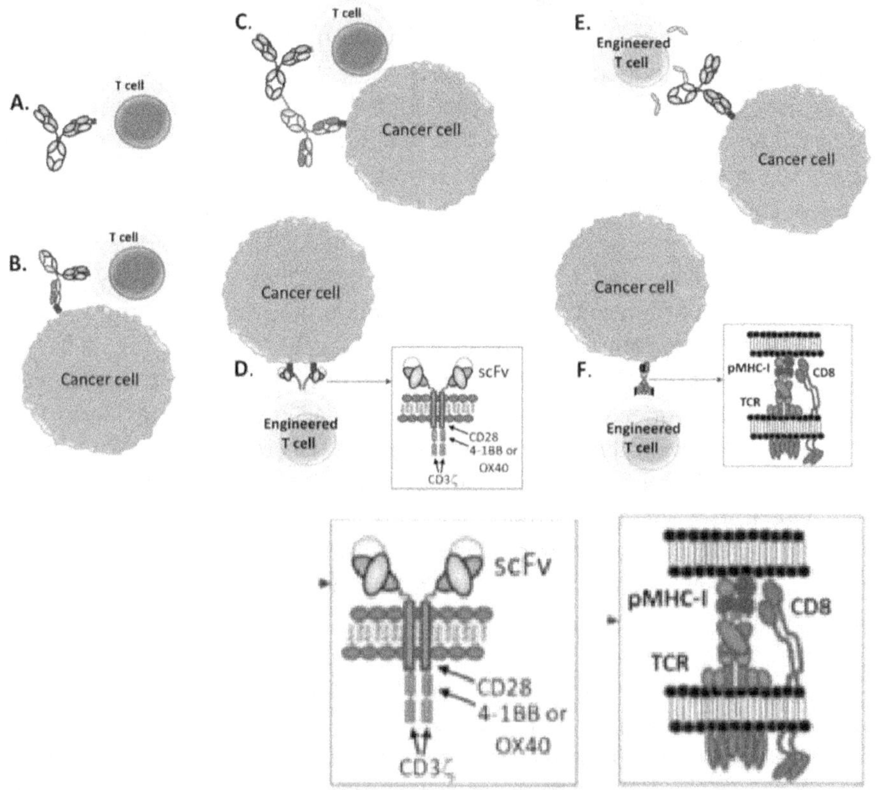

Examples of T-cell based therapeutics in clinical development. (A) Inhibition of checkpoint receptors such as PD-1 and CTLA-4 to improve T-cell activity; (B) T-cell redirection with bispecific antibodies (TRBAs) in which one binding arm recognizes a tumor antigen and the other binding arm recognizes CD3ε on T-cells; (C) Autologous T-cells activated ex vivo, combined with bispecific antibody conjugates recognizing tumor antigen with one mAb and CD3ε on T cells with the other mAb, followed by re-administration to the patient to kill tumors; (D) Genetically engineered autologous chimeric antigen receptor (CAR)-T cells in which an antibody, typically a single chain variable fragment (scFv), fused to intracellular T-cell activation domains such as CD28, 4-1BB, OX40 and CD3ζ, replace the function of the T-cell receptor (TCR), making the T-cells killers of specific antigen-bearing cells; (E) Autologous or allogeneic T-cells or NK cells genetically engineered with FcγRIIIa (CD16a), which, when administered with an anti-tumor monoclonal antibody (mAb) such as the anti-CD20 mAb, rituximab, binds to the Fc of the antibodies and functionally redirects the T-or NK-cells to the tumor to kill the cancer cells; (F). Autologous T cells with engineered TCRs. Figure and caption from Strohl and Naso.[82]
https://creativecommons.org/licenses/by/4.0/ Changed to greyscale.

29.2. *CAR-T Therapy (Chimeric antigen receptor and T for T cell)*

As a follow on to the previous figure, CAR-T therapies use engineered T-cells extracted from a patient's own blood or from an engineered generic stem-cell precursor cell. According to the "cancer.gov" website: "...over the past several years, immunotherapy-therapies that enlist and strengthen the power of a patient's immune system to attack tumors-has emerged as what many in the cancer community now call the 'fifth pillar' of cancer treatment."

The process of forming a CAR-T therapy is given below. Previously there was some background on both T cells and immune mechanisms that may help to frame these more complex concepts.

CAR T cells are the equivalent of "giving patients a living drug," explained Renier J. Brentjens, M.D., Ph.D., of Memorial Sloan Kettering Cancer Center in New York, another early leader in the CAR T-cell field. As its name implies, the backbone of CAR T-cell therapy is T cells, which are often called the workhorses of the immune system because of their critical role in orchestrating the immune response and killing cells infected by pathogens. The therapy requires drawing blood from patients and separating out the T cells.

Next, using a disarmed virus, the T cells are genetically engineered to produce receptors on their surface called chimeric antigen receptors, or CARs. These receptors are "synthetic molecules, they don't exist naturally," explained Carl June, M.D., of the University of Pennsylvania Abramson Cancer Center, during a recent presentation on CAR T cells at the National Institutes of Health campus. Dr. June has led a series of CAR T cell clinical trials, largely in patients with leukemia.

These special receptors allow the T cells to recognize and attach to a specific protein, or antigen, on tumor cells. The CAR T cell therapies furthest along in development target an antigen found on B cells called CD19 (see the box below, titled "The Making of a CAR T Cell").

Once the collected T cells have been engineered to express the antigen-specific CAR, they are "expanded" in the laboratory into the hundreds of millions.

The final step is the infusion of the CAR T cells into the patient (which is preceded by a "lymphodepleting" chemotherapy regimen). If all goes as planned, the engineered cells further multiply in the patient's body and, with guidance from their engineered receptor, recognize and kill cancer cells that harbor the antigen on their surfaces.

You can go to the cancer.gov website to follow up:
https://www.cancer.gov/about-cancer/treatment/research/car-t-cells.
A general overview graphic (modified from original) is shown below.

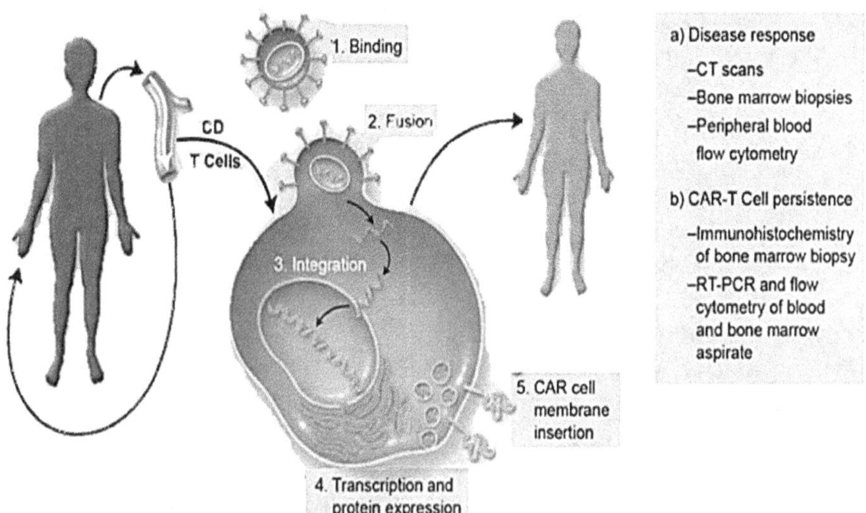

CAR-T versions include **allogenic** (key on the "all") and **autologous** (key on the "auto"). These two different classes means that either a generic version has been developed to manufacture and treat everyone (all) for off-the-shelf use or another that has been developed that depends upon modifying the patient's own T cells (auto) which means they must be removed, modified, and grown up and re-infused.

29.3. CD47–"Don't eat me" surface marker and competitive commercial development efforts

A look at the CD47 surface marker development gives a good overview of how the biotech industry searches for new blockbusters that eventually often end up at big pharma. CD47 is a so-called *"don't eat me"* signaling molecule. One company that has two development CD47 compounds is a Canadian company called Trillium (TRIL). Another is a US company called, Forty Seven. Forty Seven was recently purchased by Gilead.

The companies targeting CD47 as per a Trillium presentation publication (accessed April 2020) from their website include:

- **Trillium** (with TTI-621 and TTI-622)
- **Forty Seven** with Magrolimab (Forty Seven purchased by Gilead)
- **ALX Oncology** with ALX148
- **Celgene** (purchased by Bristol Myers) with CC-90002

- **Arch Oncology** with AD-176
- **TG Therapeutics** with TG-1801

The presentation table is especially helpful as it also lists the following differentiators for each **CD47 drug candidate:**

- **Molecule**: fusion protein or mAb or Bi-specific antibody.
- **FC isotype:** IgG1 or IgG2 or IgG4
- **RBC binding** (red blood cell binding): Yes or no with RBC binding suggested as non-desirable.
- **Monotherapy CRs**: presumably only Trillium's TTI-621 can be used as alone with no data for Arch or TG therapeutics.
- **First in human date**: given this is a race, the leader is usually the FIH.
- **Development stage:** Phase 1 a or b or P2. Only Forty Seven is in P2 as of the date of the Trillium presentation publication.

Knowing the entire CD47 ("Don't eat me") landscape is helpful but a difficulty remains: which one(s), if any, will find success? The surface marker dynamic is further expanded to both "eat me" and "don't eat me" markers as shown below derived from Wiersma et al.[83]

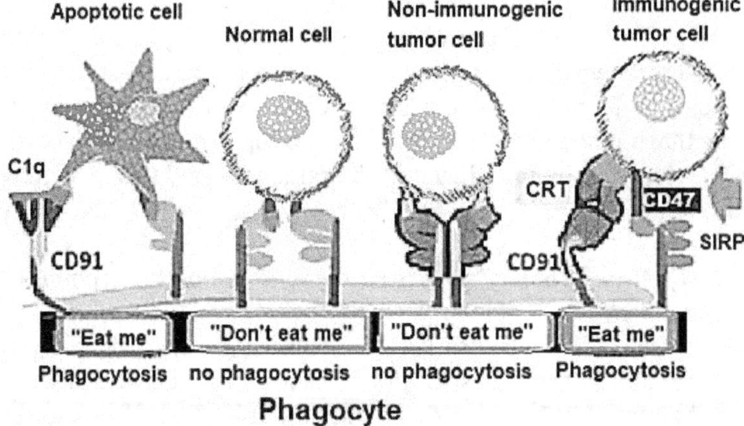

Disruption of the "don't eat me" signal. Cells express the "don't eat me" signaling molecule CD47 on their cell surface that interacts with SIRPα on phagocytes. Derived from Wiersma et al.

Above left, the complement system (C1q) makes use of "eat me" tags to bring about the phagocytosis of unwanted materials (cells and PAMPs) as pieces of complement proteins are floating in the blood constantly and as such are ready to bind and thus mark some specific PAMPs for phagocytosis.

It might not have been made clear yet, but the purpose of binding the "don't eat me" signal is that cancer cells seem to use CD47 to keep the immune system from attacking them. From the Trillium presentation (January 2020):

- Tumors use CD47 "do not eat" signals to evade destruction by innate immune system
- Many hematologic and solid tumors express high levels of CD47
- High CD47 expression correlated with aggressive disease & poor outcomes
- CD47 delivers an inhibitory "do not eat" signal to macrophages through SIRPα
- CD47 blockade emerging as a next-generation checkpoint inhibitor strategy in immune-oncology

Another emerging area of research in the search for cancer therapeutics is TIGIT as a new immune checkpoint inhibitor.

> T cell immunoglobulin and ITIM domain (TIGIT) is an inhibitory receptor expressed on lymphocytes that was recently propelled under the spotlight as a major emerging target in cancer immunotherapy. TIGIT interacts with CD155 expressed on antigen-presenting cells or tumour cells to downregulate T cell and natural killer (NK) cell functions. TIGIT has emerged as a key inhibitor of anti-tumour responses that can hinder multiple steps of the cancer immunity cycle.[84]

30. If the sky is the limit in biotechnology, what is the sky?

The sky would be to cure all disease at the genetic level. This is being attempted on several different levels with some success. The first level is (i) the continued modification of natural molecules to make entirely new entities that can solve therapeutic conundrums as shown in the graphic below. A second is (ii) modifying the central dogma of transcription by interrupting the production of protein from mRNA using "anti-sense" structures. Thirdly, the ultimate "sky is the limit" cure for disease is to (iii) replace a malfunctioning gene with a functional one.

30.1. Modification of natural molecules to make entirely new entities

Some of the more advanced current efforts to expand an already high level of biotechnology center around using functional "pieces" of antibodies rather than the entire structure for additional therapeutic benefits or reduced side effects.

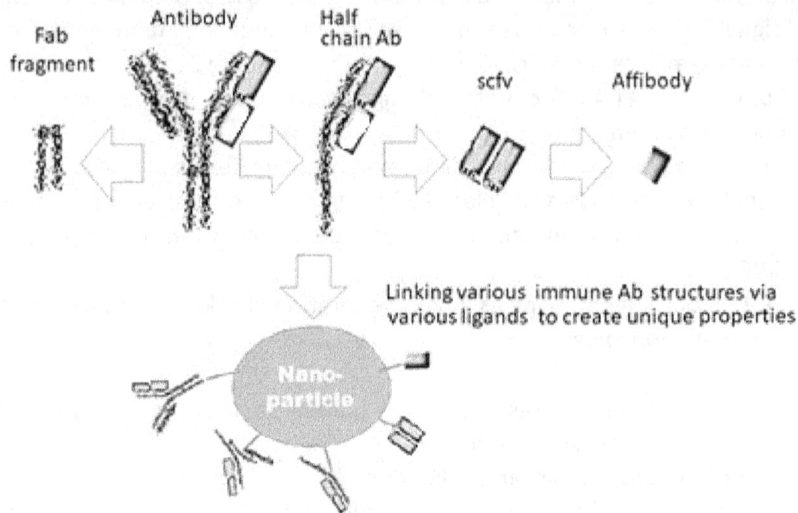

The brave new world of creating immune modulating therapies via manipulating the "natural" monoclonal antibody (top) and via linking to nanoparticles (bottom). Loosely derived from Mazzucchelli et al.[85]

Other miscellaneous biologics related treatments include anti-sense and gene therapy treatments to treat serious diseases that have no current cure including neurological diseases and inherited genetic diseases.

30.2. *Anti-sense*

Zolgensma® is a gene therapy drug that can be given to those 2 years old or less to "cure" spinal muscular atrophy. One non-gene therapy competitor to Zolgensma® is an anti-sense drug, SPINRAZA® indicated for spinal muscular atrophy developed by Ionis and Biogen. According to the manufacturer (https://www.spinraza.com/): "SMA is a rare neuromuscular disease that is characterized by the degeneration of motor neurons in the spinal cord and brainstem, leading to skeletal muscle atrophy and general weakness." SPINRAZA has gained rapid uptake and therefore generates a significant amount of revenue (>$1 billion) for Biogen and Ionis.

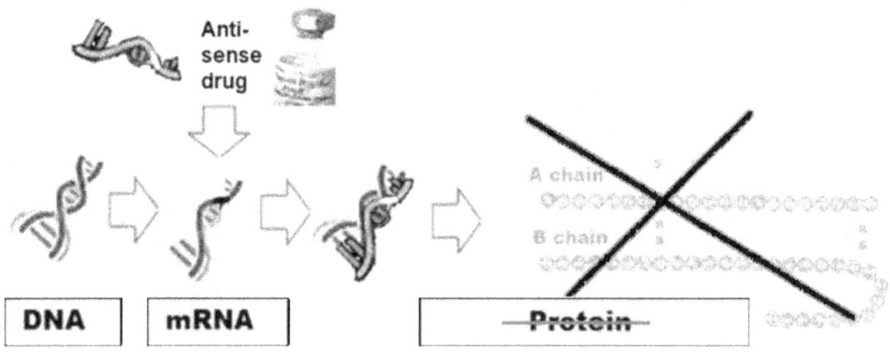

As these examples show, it is important to "Big Pharma" innovation to partner with individual biotechnology companies that have specific capabilities that BP has yet to develop.

30.3. Gene therapy

Even if treatments are discovered that are externally applied, such as the "anti-sense" therapy, eventually the "cure" will involve gene therapy application that changes the genetic problem within the patient's own DNA. However, one can expect that external improvements, even vast improvements will come sooner than eventual complete cures via gene therapy (GT). The issues to be dealt with after such cures are available will be *"when?"* and *"how much will it cost?"* and *"can patients pay?"* And *"how much will insurance pay?"*

At this point the difference between somatic and germ line encoded gene therapy should be described. Repairing a gene at the stage before development into a fetus would allow for all further cells to contain the "fixed" gene, however, thus far, fixing an errant gene in a germ line cell (a sperm cell or egg) has not been possible and in some countries is outlawed for fear of unknown risks to future generations.

GT is very promising but thus far the price tag of the solution has struck some as prohibitive. Since it is typically a once-in a lifetime injection (or at least a few time injection), then the price tag is not spread out over time but comes due all at once. This dilemma of cost is described by Financial Times (FT.com *"Biotech companies defend prices of one-off gene therapy"*), here paraphrased due to copyright prohibitions: Luxturna® is a one-time gene therapy for retinal eye disease and costs $850,000 in the US. Zolgensma® is a treatment for spinal muscular atrophy and is priced at over $2 million dollars. Biotechs have defended the drug price tags by arguing that the alternatives are more expensive when the lifetime costs of traditional drugs are tabulated against non-gene therapy drugs.

One basic method developed to get modified genes into human cells involves the use of inactivated viruses as shown below.

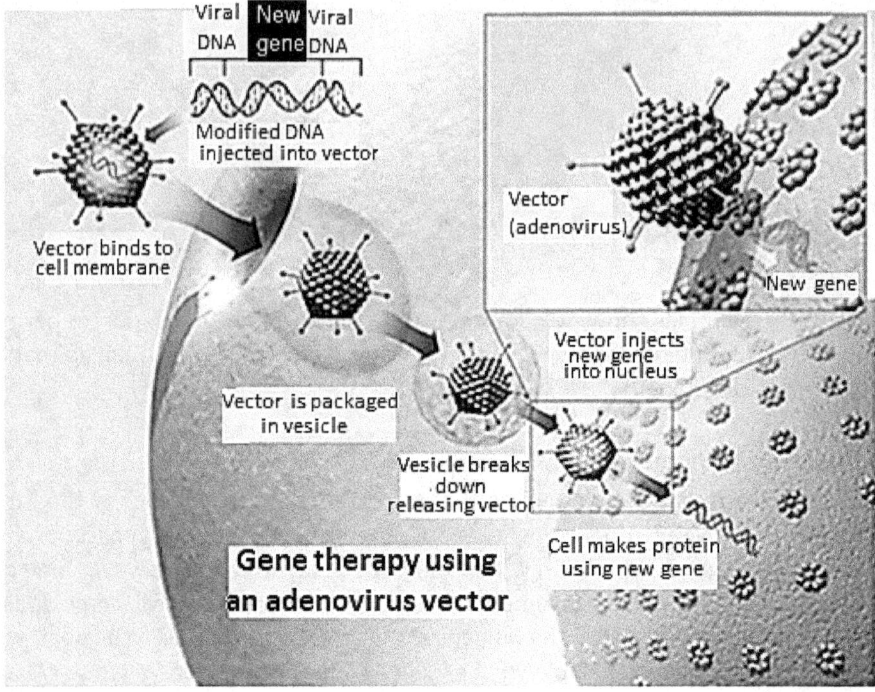

Gene therapy using an **adenovirus** vector. In some cases, the adenovirus will insert the new gene into a cell. If the treatment is successful, the new gene will make a functional **protein** to treat a disease. From NIH (captions replaced to aid legibility).

30.4. *Monster indications are difficult disease targets*
Monster indications have historically been impenetrable to cure and include devastating neurological diseases such as Huntington's (where CAG repeats accumulate from the associated faulty protein gene), Amyl lateral sclerosis (ALS), Parkinson's, Alzheimer's (with associated brain deterioration shown below), and various other nervous system related dysfunction. These potential treatments can be considered as a biotech company's "holy grail", as whoever penetrates this thick wall will open a flood gate of profits. However, given the historical difficulty of doing so makes an expectation of such breakthroughs any time soon seemingly unlikely.

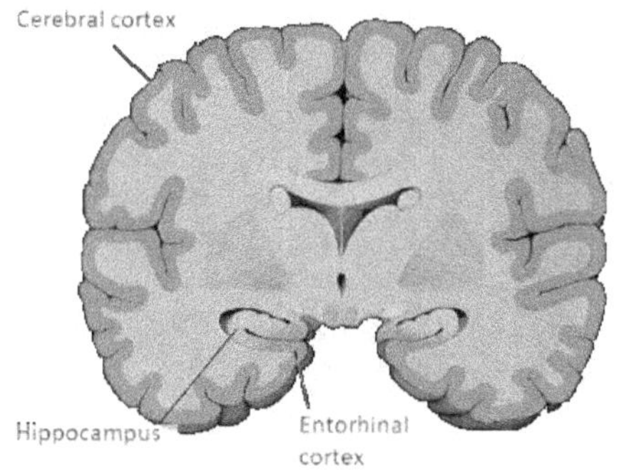

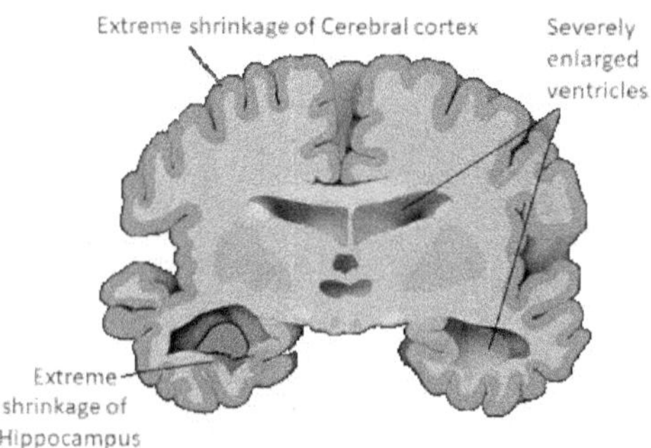

One company out in front of neurological *"monster indications"* is Biogen. With big investments in both "anti-sense" technology and "gene therapy", specifically as they have partnered with two promising smaller biotech company technologies: Ionis Pharmaceuticals (IONS) and Sangamo Therapeutics (SGMO) respectively (anti-sense and gene therapy respectively). Ionis also owns a majority stake in Akcea Therapeutics whom serves as their manufacturing arm in some cases (for some compounds) and with whom they share revenue. A significant compound Ionis-Akcea are working on is an anti-sense treatment for high cholesterol/HDL that has been shown early on to highly reduce cholesterol levels well beyond what's currently achievable.

The generally accepted mechanism of Alzheimer's progression is that the body's production of "Tau" protein aggregates interfere with the cohesion of neuron microtubules and thus as they deteriorate so does the patient's cognition and physical control. However, the formation of Tau begs the question of "how" and "why" it is accumulating and aggregating and various theories abound. Ionis/Biogen are working on an anti-sense method, so if simply chopping down the formation of Tau via complementary strands of mRNA limit the formation of Tau and this carries over to improvements or slowing of the disease process, then this will be an important milestone of at least one valid theory of Tau as a "cause" of Alzheimer's.

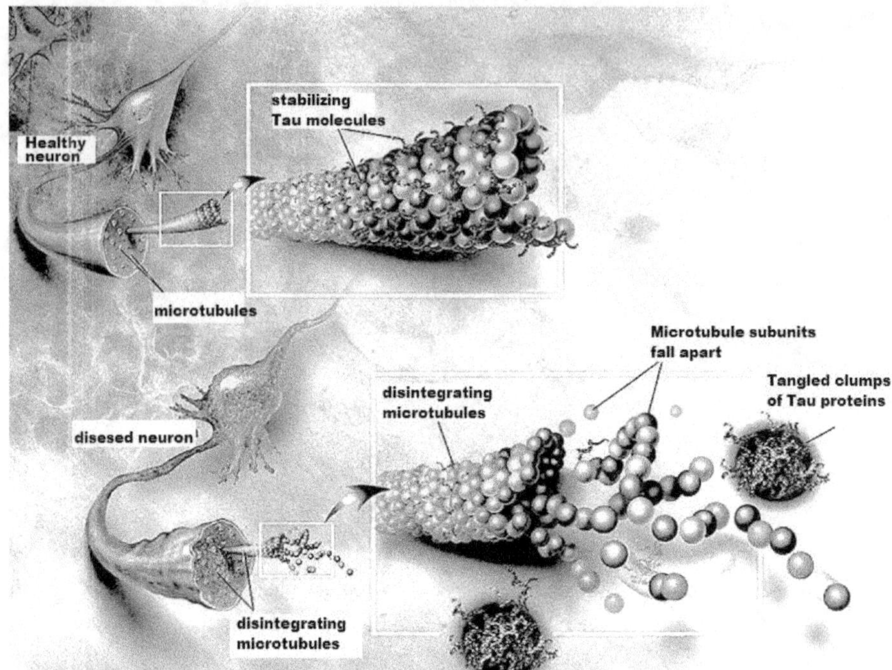

In Alzheimer's disease, changes in tau protein lead to the disintegration of microtubules in brain cells. This figure and previous figure are from: Alzheimer's Disease Education and Referral Center, a service of the National Institute on Aging (public domain).

Finally there are some early efforts (eg. Brainstorm Cell Therapeutics Inc.) to employ autologous stem cell therapy using a patient's own bone marrow to isolate and grow up Mesenchymal Stem Cells (MSCs). These MSCs are then injected into the spine to attempt totreat various destructive neurological diseases including ALS, MS and Parkinson's disease. The hope is that whatever is missing or damaged by the disease (perhaps via protein aggregation) can be supplied externally to slow disease progression. If it bears out, this is a fascinating example of using pre-diseased stem cells from a patent's own bone marrow to treat diseased derived from differentiated cells.

Glossary
Key word / definition

___

Accessory molecule-molecules used in conjunction with TLRs to detect PAMPs.

Adaptive immunity-immune response that adapts lymphocyte receptors by re-arranging them to allow detection of a myriad of structures, thus allowing a dynamic rather than static response (innate).

Amino acids-the subunits or building blocks that make up proteins. There are 20 different amino acids incorporated into human proteins.

Antibody- an immunoglobulin (Ig), a large, Y-shaped protein produced mainly by plasma cells (differentiated from B cells) that are used to neutralize pathogens by binding to antigens and being devoured via phagocytes.

Antibody diversity-the variable regions of antibodies are generated by somatic rearrangement of germline encoded set of genes such that the potential diversity of structures is almost infinite.

Antigen-particle or surface structure that antibody binds to. Antigen is to antibody as PAMP is to PRR.

Antisense-complementary strand of mRNA to the coding mRNA as used therapeutically to cancel prospective destructive protein production.

APC, antigen presenting cell- or accessory cells (macrophage, B cells, dendritic) that display antigen complexed with major histocompatibility complexes (MHCs) on their surfaces; a process known as antigen presentation.

B cell-a white blood cell that is a lymphocyte that secrete B cells, cytokines, and present antigen. They mature in the "bursa of Fabricus" of birds and bone marrow of mammals.

Bacteriophage - is a virus that infects and replicates within bacteria and archaea.

BCR, B cell receptor- is a (type 1) transmembrane protein on the surface of a B cell. B cell receptors are composed of immunoglobulin molecules, and are located on the outer surface of these lymphocyte cells but also correspond to the type binding regions found on the secreted antibodies from the same cell.

Biologic- old and new drugs derived from living organisms via natural or recombinant production.

CAR-T, chimeric antigen receptor T cells- genetically engineered T cells created to produce an artificial T-cell receptor for use in immunotherapy. This "artificiality" is designed to bind specific proteins. They are chimeric because they combine both antigen-binding and T-cell activating functions into a single receptor.

CD marker- cluster of differentiation (or designation or classification determinant) abbreviated as CD. It is used for the identification and investigation of cell surface molecules providing targets for immunophenotyping of cells.

CDR, complementary determining region- are part of the variable chains in immunoglobulins (antibodies) and T cell receptors, generated by lymphocytes (B-cells and T-cells respectively), where these molecules bind to their specific antigen. A set of CDRs constitutes a paratope (and pairs with or detects the epitope of the antigen).

Cell culture- a tool of biotechnology where cells are grown without a host organism for the purpose of producing natural or recombinant proteins.

Central tolerance-the mammalian immune system's way of learning what proteins are self and to therefore later distinguish which proteins are non-self.

Coagulation- a metazoan tool to isolate and facilitate removal of contaminants (I.e. bacterial invaders).

Complement-protein and peptide particles, often proteases, that attach to bacteria and other cells to mark them for phagocytosis.

Cytokine- broad category of small proteins (~5–20 kDa) important in cell signaling. Cytokines are peptides and cannot cross the lipid bilayer of cells to enter the cytoplasm. Cytokines have been shown to act as immunomodulating agents to facilitate inflammation.

Cytometry- is the measurement of the characteristics of immune cells, especially surface markers to differentiate one immune cell type from another.

Deuterostome- in Greek means "second mouth" and constitutes a superphylum of animals. It is a sister clade to Protostomia (Protostomes).

Drosophila- a genus of small fruit flies used as a model organism for genetic studies due to possessing large chromosomes and speedy reproduction.

Effector- a small molecule that selectively binds to a protein and regulates its biological activity. Effector molecules, therefore, act as ligands that can increase or decrease enzyme activity, gene expression, or cell signaling.

Endosymbiotic theory- The theory that eukaryotic cells contain once free-living prokaryotic cells as organs including mitochondria, plastids and chloroplasts.

Endotoxin or lipopolysaccharide-a major outer surface constituent of Gram negative bacteria and potent PAMP for metazoans.

Eukaryote- Eukaryotic cells are cells that contain a nucleus and organelles, and are enclosed by a plasma membrane. Includes protists, fungi, plants and animals.

Exotoxin- secreted microbial toxins that act on competitor and host organisms are small molecule proteins and include superantigens.

Exon- sections of a Eukaryotic gene that are spliced out leaving the expression of the introns that produce the functional protein.

Germ theory- the early recognition that microorganisms are responsible for disease, food spoilage and infection.

Gram stain- a method of staining that reveals the dichotomy of Gram positive (single thick membrane consisting mostly of peptidoglycan) and Gram negative organisms [a double outer membrane consisting largely of LPS, phospholipids, and protein (porins)].

HLA, human leukocyte antigen- system or complex is a group of related proteins that are encoded by the major histocompatibility complex (MHC) gene complex in humans. The proteins encoded by certain genes are also known as *antigens*, as a result of their historic discovery as factors in organ transplants. Different classes have different functions including, MHC I and MHC II which display intrinsic and extrinsic proteins on the surface for immune surveillance respectively.

Hox genes- a group of related genes that specify regions of the body plan of an embryo along the head-tail (dorso-ventral) axis of animals. Hox proteins encode and specify the characteristics of 'position', defining "what goes where". These genes have been conserved from antiquity (as have other fundamental mechanisms including DNA, RNA, ribosomes, porins, etc.).

Inflammation- complex response of body to harmful stimuli, such as pathogens, damaged cells, or irritants, and involves immune cells, blood vessels, and molecular mediators.

Innate immunity- static, germ-line encoded, mediators of first line immune defense.

Intron- sections of a Eukaryotic gene that encode a protein after the exons are spliced.

Ligand- a substance that forms a complex with a biomolecule to serve a biological purpose, for example, endotoxin is a ligand of MD-2 and TLR4.

Limulus- the genus of horseshoe crab with *Limulus polyphemus*, being one of four remaining (extant) horseshoe crab species. Used as a model organism for study of its blood and optical systems.

Lipopolysaccharide- endotoxin, a major PAMP from Gram negative bacteria that is a potent activator of the immune system of most animals.

LRR, leucine rich repeat- is a protein structural motif that forms an α/β horseshoe fold. It is composed of 20–30 repeating amino acid stretches that are rich in the hydrophobic amino acid leucine. These are used in various forms to detect/capture PAMPs such as in Toll and TLRs.

Metazoan- multicellular organisms

MHC-1- a carrier structure for internal proteins that are carried to the surface and surveilled by T cells for self or non-self. If determined to be foe, the cell it is presented by is attacked by a cytotoxic T cell. In humans, the HLAs corresponding to MHC class I are HLA-A, HLA-B, and HLA-C.

MHC-2- molecules normally found only on professional APCs such as dendritic cells, mononuclear phagocytes, some endothelial cells, thymic epithelial cells, and B cells. The antigens presented by class II peptides are derived from extracellular proteins (via phagocytosis) rather than from internal proteins as is done with MHC-1.

Monoclonal antibody- a product of biotechnology where the antibodies are made by the culture expansion of identical immune cells which are all clones belonging to a unique parent cell.

PAMP, pathogen associated molecular pattern- pieces or parts or secreted proteins of microbes (bacteria, fungi, protozoa) that serve to metazoans of immune invasion.

Peptide- a small chain of amino acids that is not quite the size of a protein.

Phagocytosis- engulfment of singular celled organism by another or by a metazoan specialized immune cell.

Porin- the "holes" (trimeric proteins) in cells that allow for passage of food-stuff and gases.

Prion- infectious protein particle that can be activated via a domino effect from an initial particle.

Prokaryote- single celled organisms including bacteria, that lacks an envelope-enclosed nucleus. From the Greek (*pro*, 'before' and *karyon*, 'nut' or 'kernel').

Protease- an enzyme that catalyzes the breakdown of proteins. Blood proteins often exist as zymogen forms that when activated begin to set off a chain of reactions to quickly mount coagulation as an immune response.

Protostome- a clade of animals containing phyla including the arthropods, annelids, and mollusks that embryonically form "mouth first".

Protein- a functional string of amino acids that are the product of a gene production.

PRR, pattern recognition receptor- germline-encoded host immune cell receptors, which detect molecules typical for the pathogens (PAMPs) and for host tissue breakdown byproducts (DAMPs).

Receptor-protein structures that receive and transduce signals that may be integrated into biological systems. These signals are typically chemical messengers which bind to a receptor and cause a cellular/tissue response.

Recombinant protein- a product of biotechnology that allows for the mass production via cell culture of natural or genetically unique proteins for therapeutic and other purposes.

<u>Restriction endonuclease</u>- an enzyme derived from bacteria or virions that cleave DNA into fragments at or near specific recognition sites within molecules known as restriction sites.

<u>Sepsis</u>- a harmful imbalance of immune non-homeostasis brought about by infection or artifacts from previous infection.

<u>Superantigen</u>- a class of antigens from bacterial and viral sources (including exotoxins) that result in excessive activation of the immune system. SAgs cause non-specific activation of T-cells resulting in polyclonal T cell activation and massive cytokine release.

<u>T cell</u>- a type of lymphocyte that develops in the thymus.

<u>TCR, T cell receptor</u>- a receptor on T cells that that is responsible for recognizing fragments of antigen as peptides bound to binds MHC-1 or MHC-2.

<u>Toll</u>- a *Drosophila* innate immune receptor (a PRR) that serves as A PAMP detector in the fly. Toll lead to the discovery of Toll-Like Receptors as transmembrane structures conserved from early animals.

<u>Vaccination</u>- the use of a disease-causing substance (i.e. protein) in a weakened state (killed or inactivated or a partial structure) to bring about full immunity to an associated disease in a prophylactic manner.

Blank page

References

1 Bruce A. Beutler, Nobel Lecture, How Mammals Sense Infection: From Endotoxin to the Toll-like Receptors, The Nobel Prize in Physiology or Medicine 2011.

2 The momentous transition to multicellular life may not have been so hard after all, Elizabeth Pennisi, Sciencemag.org, Jun. 28, 2018 , 12:30 PM.

3 Brandon Keim, 2-Billion-Year-Old Fossils May Be Earliest Known Multicellular Life, Science, June, 30, 2010, 03:24 PM.

4 Evolution of Invertebrate Deuterostomes and Hox/ParaHox Genes, Tetsuro Ikut, Genomics Proteomics Bioinformatics 2011 Jun; 9(3): 77-96.

5 Lutz B, et al. Rescue of Drosophila Labial Null mutant by the chicken ortholog Hoxb-1 demonstrates that the function of Hox genes is phylogenetically conserved, Genes Dev., 1996;10:176–84.

6 Improving Hox Protein Classification across the Major Model Organisms, Stefanie D. Hueber et al., PLoS ONE, www.plosone.org 1, May 2010, Volume 5, Issue 5, e10820.

7 Antonie van Leeuwenhock letter to Royal Society, Delft. Lampham's Quarterly. 2018;XL(3): 65. Summer.

8 Rietschel ET, Cavaillon M. Richard Pfeiffer and Alexandre Besredka: creators of the concept of endotoxin and anti-endotoxin. Microbes Infect. 2003;5:1407–14.

9 High-resolution architecture of the outer membrane of the Gram-negative bacteria Roseobacter denitrificans, Jarosławski et al., Mol Microbiol. 2009 Dec;74(5):1211-22.

10 Liu C, et al. Immunogenic characterization of outer membrane porins OmpC and OmpF of porcine extra-intestinal pathogenic *Escherichia coli*, FEMS Microbiol Lett. 2012;337:104–11.

11 Porins in prokaryotes and eukaryotes: common themes and variations, Zeth and Thein, Biochem J. 2010;431:13–22.

12 The 3D structures of VDAC represent a native conformation, Hiller et al., Trends in Biochemical Sciences 35 (2010) 514–521.

13 Pirmoradian et al., Isoelectric point region pI≈7.4 as a treasure island of abnormal proteoforms in blood. Discoveries. 2016;4(4):e67.

14 Antimicrobial Peptides from Marine Invertebrates, Tincu and Taylor, Antimicrob Agents Chemother, 2004 Oct; 48(10): 3645–3654.

15 Antimicrobial Peptides, Isolated from Horseshoe Crab Hemocytes, Tachyplesin II, and Polyphemusins I and II: Chemical Structures and Biological Activity, Toshiyuki Miyata, *The Journal of Biochemistry*, Volume 106, Issue 4, October 1989, Pages 663–668.

16 Antimicrobial Peptides from Marine Invertebrates, Tincu and Taylor, Antimicrobial Agents and Chemotherapy, Oct. 2004, p. 3645–3654.

[17] Kim et al., An Insect Multiligand Recognition Protein Functions as Opsonin for the Phagocytosis of Microorganisms, The Jour. Of Biol. Chem., Vol. 285, NO. 33, pp. 25243–25250, August 13, 2010.

[18] Model Systems of Invertebrate Allorecognition, Rosengarten and Nicotra, Current Biology, 21, R82-R92, Jan. 25, 2011.

[19] Genomic basis of evolutionary change: evolving immunity, Bregje Wertheim, Frontiers in Genetics, Jun. 2015, Vol. 6, article 222.

[20] O'Neill, Golenbock and Bowie, The history of Toll-like receptors- redefining innate immunity, Nature Reviews Immunology, 2013.

[21] Lemaitre et al. The dorsoventral regulatory gene cassette *spätzel/Toll/cactus* controls the potent antifungal response in Drosophila adults, Cell, 86: 973-983.

[22] Vogel, How discovery of Toll-mediated innate immunity in Drosophila impacted our understanding of TLR signaling (and vice versa), Jour. Immunology,

[23] Bruce Beutler, Xin Du, Alexander Poltorak, Identification of Toll-like receptor 4 (Tlr4) as the sole conduit for LPS signal transduction: genetic and evolutionary studies, Journal of Endotoxin Research, Vol. 7, No. 4, 2001.

[24] Janeway, Approaching the asymptote? Evolution and revolution in immunology, Cold Spring Harb Symp Quant Biol. 1989;54 Pt 1:1-13.

[25] Nothing in Biology Makes Sense except in the Light of Evolution, Theodosius Dobzhansky, The American Biology Teacher, Vol. 35, No. 3 (Mar., 1973), pp. 125-129 Published by: National Association of Biology.

[26] Bier E, McGinnis W. Chapter 3, Model organisms in the study of development and disease. In: Erickson RP, Wynshaw-Boris AJ, editors. Epstein's inborn errors of development: the molecular basis of clinical disorders of morphogenesis (Oxford Monographs on Medical Genetics) 3rd ed. Oxford University Press; 2016

[27] Muta et al., Limulus Factor C: an endotoxin-sensitive serine protease zymogen with a mosaic structure of complement-like, epidermal growth factor-like, and lectin-like domains, Jour. Biol. Chem., Vol. 266, No. 10, Issue of April 5, pp. 6554-6561, 1991.

[28] Profiling the Enzymatic Properties and Inhibition of Human Complement Factor B, Giang Thanh Le et al. J. Biol. Chem. 2007;282:34809-34816.

[29] Sharma, Chelicerates and the conquest of land: a view of arachnid origins through an evodevo spyglass. Integr Comp Biol. 2017;57(3):510–22. https://doi.org/10.1093/icb/icx078.

[30] Damen WG, et al. Diverse adaptations of an ancestral gill: a common evolutionary origin for wings, breathing organs, and spinnerets. Curr Biol. 2002;12:1711–6.

[31] Levin, J. and Bang, F.B., Clottable protein in Limulus: its localization and kinetics of its coagulation by endotoxin, Thrornb. Diath.Haemorrh., 1968, 19:186-197.

[32] Zhang D, et al. A novel molecule designed to keep bacteria out of the urinary tract a Toll-like receptor that prevents infection by uropathogenic bacteria. Science. 2004;303:1522–6.

[33] Leucine-rich repeat (LRR) proteins Integrators of pattern recognition and signaling in immunity, Aylwin Ng and Ramnik J., Xavier Center for Computational and Integrative Biology, and Gastrointestinal Unit; Massachusetts General Hospital; Harvard Medical School; and Broad Institute of MIT and Harvard; Cambridge, MA USA. Autophagy 7:9, 1082-1084; September 2011; © 2011 Landes Bioscience

[34] Robinson and Moehle, Structural aspects of molecular recognition in the immune system, Part II: Pattern recognition receptors, Pure and Applied Chemistry, 86(10):1483-1538.

[35] Merle NS, et al. Complement system part II: role in immunity. Front Immunol. 2015;6:257.

[36] Ramirez-Ortiz and Means. The role of dendritic cells in the innate recognition of pathogenic fungi (A. fumigatus, C. neoformans and C. albicans). Virulence. 2012;3(7):635–46.

[37] Roitt et al., Routt's Essential Immunology, 2019, 13th edition, 2017, Wiley Blackwell.

[38] Monie, Tom P., The Innate Immune System, 2017, Academic Press.

[39] Beutler B, Poltorak A. The sole gateway to the endotoxin response: how LPS was identified as TLR4, and its role in innate immunity. Drug Metab Dispos, The American Society for Pharmacology and Experimental Therapeutics. 2001;29(4):Part 2.

[40] The molecular mechanism of species-specific recognition of lipopolysaccharides by the MD-2/TLR4 receptor complex, Alja Oblak and Roman Jerala, Molecular Immunology, Volume 63, Issue 2, February 2015, Pages 134-142.

[41] Acha-Orbea, Bacterial and viral superantigens: roles in autoimmunity?, Ann Rheum Dis. 1993 Mar; 52(Suppl 1): S6–16.

[42] Salzet M. Vertebrate innate immunity resembles a mosaic of invertebrate immune responses. Trends Immunol. 2001;22(6):285.

[43] Koshiba T, Hashii T, Kawabata S., A structural perspective on the interaction between lipopolysaccharide and factor C, a receptor involved in recognition of Gram-negative bacteria, Jour. Biol Chem., 2007;282:3962–7.

[44] Stanley Prusiner, Novel proteinaceous infectious particles cause scrapie, Science. 1982 Apr 9;216(4542):136-44.

[45] Watts et al., The expanding universe of prion diseases. PLoS Pathology. 2006;2(3):e26.

[46] Rast and Buckley, Lamprey immunity is far from primitive, PNAS, April 9, 2013, vol. 110, no. 15, pages 5746–5747.

[47] Ronald N. Germain, M.D., The Cellular Determinants of Adaptive immunity Immunity, New England Jour. Med., 381;11, September 12, 2019.

[48] Cooper MD, Peterson RD, Good RA. Delineation of the thymic and bursal lymphoid systems in the chicken. Nature 1965; 205:143-6.

[49] Cooper MD, Raymond DA, Peterson RD, South MA, Good RA. The functions of the thymus system and the bursa system in the chicken. J Exp Med 1966;123:75-102.

[50] Miller JFAP, Mitchell GF. Cell to cell interaction in the immune response. I. Hemolysin-forming cells in neonatally thymectomized mice reconstituted with thymus or thoracic duct lymphocytes. J Exp Med 1968;128:801-20.

[51] Templeton and Moehle, Structural aspects of molecular recognition in the immune system. Part I: Acquired immunity, Pure and Applied Chemistry, 2014; 86(10): 1435-1481.

[52] Antigen phagocytosis by B cells is required for a potent humoral response, Martínez-Riaño et al., EMBO reports 19: e46016, 2018.

[53] Brandt and Roth, G.O.D.'s Holy Grail: Discovery of the RAG Proteins, J Immunol 2008; 180:3-4.

[54] Tonegawa S. Nobel Prize in Physiology or Medicine 1987. Available online at: https://www.nobelprize.org/prizes/medicine/1987/summary/ (accessed March 20, 2019).

[55] Immunology's Coming of Age, Stefan H. E. Kaufmann, Front. Immunol., 03 April 2019.

[56] Janeway, Travers, and Walport et al., Immunobiology: The Immune System in Health and Disease, 5th edition, New York: Garland Science; 2001.

[57] Jack and DuPasquier, Evolutionary concepts in immunology, Springer, 2019.

[58] CD28 co-stimulation in T-cell homeostasis: a recent perspective, Beyersdorf et al., ImmunoTargets and Therapy, 2015.4, 111-122.

[59] Luckheeram et al., CD4+T Cells: Differentiation and Functions, Clinical and Developmental Immunology Volume 2012, Article ID 925135.

[60] Memory B Cells of Mice and Humans, Weisel and Shlomchik, Annu. Rev. Immunol. 2017. 35:255–84.

[61] Tauber A., Immunology's theories of cognition, Hist Philos Life Sci. 2013;35(2):239-64.

[62] A trip through my life with an immunological theme, Charles A. Janeway, Jr., Annu. Rev. Immunol. 2002. 20:1–28.

63 Yanaba et al. "B-lymphocyte contributions to human autoimmune disease". *Immunological Reviews*. 223 (1): 284–299, 2008-06-01.

64 Shaffer et al., Pathogenesis of Human B Cell Lymphomas, Annual Rev. Immun. 30(1): 565-610, Jan 2012.

65 T cell epitope redundancy: cross-conservation of the TCR face between pathogens and self and its implications for vaccines and auto-immunity, Moise and Beseme, et al., Expert Review of Vaccines, 2015.

66 Co-stimulatory and co-inhibitory pathways in autoimmunity, Zhang and Vignali, Immunity 44, May 17, 2016.

67 Costimulatory pathways in transplantation, Pilat et al., Seminars in Immunology 23 (2011) 293-303.

68 A primer on recent developments in cancer immunotherapy, with a focus on neoantigen vaccines, Tokuyasu and Huang, J. Cancer Metastasis Treat, 2018,4:2.

69 Mechanisms of immune-related adverse events associated with immune checkpoint blockade: using germline genetics to develop a personalized approach, Zia Khan et al., Genome Medicine, 2019; 11: 39.

70 De-novo and acquired resistance to immune checkpoint targeting, Syn et al., Lancet Oncology, 2017 Dec;18(12):e731-e741.

71 Ganellin R, Jefferis R, Roberts S. Biological and small molecule drug research and development: theory and case studies: Academic Press/Elsevier; 2013.

72 Genentech. Biosimilars. http://www.gene.com/gene/about/views/followon-biologics.html

73 Nunes-Alves, C. "Blood is a very unusual fluid". *Nat Immunol* 17, S5 (2016). https://doi.org/10.1038/ni.3600.

74 Hsiao C, et al., Peeling the onion: ribosomes are ancient molecular fossils, Mol Biol Evol., 2009;26(11):2415–25.

75 *Berg Boyer Cohen: The invention of recombinant DNA technology*, Mark Jones, LSF Magazine, https://medium.com/lsf-magazine, Nov. 10, 2015, accessed 03/29/2020.

76 Nguyen N.Y., et al., Isolation and characterization of Limulus C-reactive protein genes, Val. 261, No. 22, Issue of August 5, pp. 10450-10455,19.

77 Bulletin of the World Health Organization, 61(6): 897-911 (1983), Quality control of biologicals produced by recombinant DNA techniques.

78 *Mihaela Pertea & Steven Salzberg (2010).* "Between a chicken and a grape: estimating the number of human genes". *Genome Biology. 11 (5): 206.* doi:10.1186/gb-2010-11-5-206. PMC 2898077. PMID 20441615.

79 *Venter, JC; et al. (2001).* "The sequence of the human genome". *Science. 291 (5507): 1304–1351.* Bibcode:2001Sci...291.1304V.

80 Bryant, J. A (2007). Design and information in biology: From molecules to systems. p. 108. ISBN 9781853128530. ...brought to light about 1200

protein families. Only 94 protein families, or 7%, appear to be vertebrate specific

[81] Gong et al., Development of PD-1 and PD-L1 inhibitors as a form of cancer immunotherapy: a comprehensive review of registration trials and future considerations, Journal for ImmunoTherapy of Cancer (2018) 6:8.

[82] Bispecific T-Cell Redirection versus Chimeric Antigen Receptor (CAR)-T Cells as Approaches to Kill Cancer Cells, Strohl and Naso, Antibodies 2019, 8, 41.

[83] Wiersma et al., Mechanisms of translocation of ER chaperones to the cell surface and immunomodulatory roles in cancer and autoimmunity, Front. Oncol. 20 Jan. 2015.

[84] Harjunpää and Guillerey, TIGIT as an emerging immune checkpoint, Clinical and Experimental Immunology, 200: 108–119

[85] Mazzucchelli S, et al. Targeted approaches for HER2 breast cancer therapy: news from nanomedicine? World J Pharmacol. 2014;3(4):72–85.